Handling
Elevator
Emergencies
Revised Edition

By
David H. Cowardin

TABLE OF CONTENTS

Acknowledgments

During the development of the first edition of this text, many individuals gave unselfishly of their time and efforts to help the author present the concepts and technical knowledge necessary to communicate complex procedures in a meaningful and understandable manner to others. Special thanks also goes to the technical editors of the first edition of this book: Many of these individuals I have lost touch with over the years, most are likely retired, and some may have pasted to the great beyond, but I still remember their contributions to the original text.

Edward Donoghue, Manager Codes and Safety, National Elevator Industry, Inc.

Jim O'Boyle, CSP Safety Director, Dover (Dover Sold to ThyssenKrupp 1999) Corporation Elevator Division

Lt. William 1. Coakley, Boston Fire Department

Gus Armbruster, Haughton Elevator (Purchased by Schindler Group 1979), whose assistance helped to make this book possible.

The late Dan W. Iler, formerly with the Palos Verdes Estates Fire Department, whose manuscript *Elevator Emergencies* was an indispensable source in writing this book.

I would also like to acknowledge the National Fire Protection Association who published the first edition of this text and who reverted the rights to that publication to me making this edition possible.

And, finally, I would like to thank the untold number of elevator installers, adjusters, and maintenance personnel who gave generously of their time to help firefighters improve their ability to handle elevator emergencies.

5

Introduction

The original text published in 1979 stated "that in the United States today, there are more than 340,000 passenger elevators in operation." Today that number is estimated to be between 600,000 and 700,000 depending on what source is used. One elevator car carries an average of four passengers on approximately ten round trips during an eight-hour working day. This means that one passenger elevator transports approximately 30,000 people per year, more than the population of many good-sized cities. It has been said that elevators carry more passengers than all the other forms of mass transportation in the United States. I don't know if that statement is true or not, but when you consider the number of elevators it is not difficult to conclude that they play a major role in our society.

Each year between seven and eight thousand high-rise building fires are reported and some 10,000 people wound up in the emergency room[1] because of elevator-related accidents even with all this elevators possess a remarkable safety record. They do however present unique problems to persons employed in emergency professions. These problems appear to be relatively unimportant before an emergency occurs, due possibly to the infrequency of incidents involving elevators. When faced with a situation where an elevator is involved in an emergency, however, rescue personnel realize they need to know how an elevator works and the procedures used for rescuing passengers or for using elevators in fire fighting.

---

[1] Good Housekeeping, "The last word on elevator safety", 1998

With an increasing number of high-rise buildings dotting our urban areas, the elevator has assumed an important role as a form of vertical transportation and given rise to an entire industry. Many companies install elevator systems and many more manufacture elevator components. Added to the variety of installations and manufacturers are local, county, and state codes and ordinances that deal directly with elevator safety devices and component parts.

Due to this variety in elevator construction, installation, and codes, there are an infinite number of situations which may confront rescue personnel. This book will present general procedures for dealing with the most commonly encountered situations involving elevators, with the purpose of familiarizing rescue personnel with elevator installations, operating characteristics, and effectual, industry-approved rescue and fire fighting methods. While no guarantee is either expressed or implied to any organization or individual who implements these procedures, they should provide rescue personnel with a basis for knowledgably handling situations involving elevators.

David H. Cowardin

CHAPTER 1

Elevator Installations

FIRE SERVICE ELEVATOR CONCERNS

Elevators travel more miles and carry more people than any other form of transportation, In *Man on the Move,* a 1967 publication by' Harvey S. Firestone, Jr.[2], estimates show that in one year alone elevators safely carried more than twenty-eight billion people *over* 500 million miles worldwide. Today, more than a 40-years later, 700,000 elevators in the U.S. alone carry approximately 100 billion passengers almost 400 million miles each year, or four times the distance to the sun. Elevators also maintain a safety record better than that of any other form of transportation.

Elevators serve virtually all types of structures, including libraries, apartment buildings, hotels, warehouses, department stores, parking garages, and hospitals. Thus, elevator emergencies are of major concern to firefighters who take part in emergency rescue operations. Firefighters are sometimes the first to arrive at the scene of an elevator emergency, and they may find themselves exposed to very dangerous situations. They may, for example, need to use elevators

---

[2] Firestone, Harvey s., Jr., Man on the Move, G.P. Putnam's Sons, NY, 1967, p. 194

to fight fires in multistory buildings. Since these buildings are likely to have express elevator's that do not have openings for many consecutive floors, firefighters must be extremely careful to avoid entrapment.

Common Causes of Elevator Emergencies

Some of the more common causes of elevator emergencies are power failures, electrical malfunctions, and stalled elevator cars. It is therefore necessary that fire service personnel have a basic understanding of elevator operation, a familiarity with elevator components, and an understanding of emergency rescue procedures if they are to handle elevator emergencies effectively. Without this knowledge, a simple procedure can become a complex operation.

It can be extremely dangerous to attempt a rescue of trapped elevator car passengers during an elevator emergency - dangerous both for firefighters and trapped passengers. Except during fires, trapped passengers are usually safer when left within a stalled elevator car. Passengers are endangered when they leave an elevator car in any way other than walking out normally.

Whenever possible, an elevator emergency should be handled with the assistance of an elevator mechanic or manufacturer's representative. The training and experience of these persons enables them to cope with the complexities of the elevator system and ensures greater safety for the passengers inside the car.

Safety Regulations

Rigid specifications and regulations are imposed on the design, construction, testing, installation, and operation of elevators. State and municipal authorities usually use ANSI/ASME A17.1[3] - Safety Code for Elevators and Escalators, published by the American Society of Mechanical Engineers, as the basis for their own elevator ordinances. Since its first edition in 1921, this code has been revised to reflect advancements in the state of elevator art. In 1971, ASME developed ANSI A17 Guide for the Evacuation of Passengers from Stalled Elevator Cars that has also been revised over the years and is now known as ANSI/ASME 17.4 Emergency Evacuation of Passengers from Elevators. The Inspection services and provisions for the safe operation of elevators are discussed later in this chapter in the section titled "Safety Provisions for Elevators."

---

[3] American National Standard Institute/American Society of Mechanical Engineers

ELEVATOR COMPONENTS

According to various model building code definitions, an elevator is a hoisting and lowering mechanism, designed to carry passengers or authorized personnel, equipped with an elevator car which moves in fixed guides and serves two or more fixed landings. However, some conveyances that move in horizontal and diagonal directions are classed as elevators. In many large cities, for example, high-rise parking garages use horizontally moving elevators or cages (mechanical parking garage equipment) to transport vehicles to and from available spaces without having to move parked cars. In addition, diagonally moving elevators are used to transport people and materials to structures built on top of cliffs or hills and move materials from ships or trains to the upper floors of storage structures. Although horizontal and diagonal elevators are not common, fire service personnel should be aware of their presence in the community and should contact manufacturers for specifications and operational information. However, because most elevators do not travel horizontally or diagonally, this text will provide information only on elevators that operate vertically.

Elevator Operation

The vertical area of elevator travel is the hoistway, variously called the hatch, hatchway, or elevator shaft. The hoistway can be used for one car (a single hoistway) or more than one car (a multiple hoistway). The hoistway area includes all space from the bottom of the pit (the area below the lowest landing of the elevator car) up~ ward to the underside of the floor or roof at the top of the hoistway. Some hoistways pass floors in a building that have no normal landing entrance: these are called blind hoistways, and are most often found in high-rise buildings that contain express elevator service.

Elevator cars are kept on a vertical path within the hoistway by means of guide rails. Guide rails help minimize the car's side sway. Roller guides or guide shoes maintain the elevator's vertical travel within the hoistway, and connect the elevator to the guide rails. The crosshead is the top beam of the car's frame, and the safety plank or bolster is the bottom of the frame. They are connected to each other by uprights known as stiles.

Generally, elevator cars are lifted and lowered either by hydraulic pressure or by electric motors. A pressurized liquid serves as the basis for operating a hydraulic elevator system. Oil, which usually serves as the pressure fluid, is supplied through a motor-driven, positive placement pump that is activated by an electric-hydraulic control system. The hoisting and lowering mechanism for the hydraulic system is a plunger (also called a ram). (See Figure 1.1.) The elevator car sits atop the plunger, which operates in a pressure cylinder. To raise the car, the pump discharges oil into the pressure cylinder, forcing the plunger up. When the car reaches the desired level, the pump is stopped. To lower the car, oil is released from the pressure cylinder and returned to a storage tank (reservoir) by a valve, easing the pressure and thereby lowering the car. A diagram of a typical hydraulic elevator is shown in Figure 1.2.

Fig. 1.1. *This is the pit area of a hydraulic elevator. The ram or plunger is in the center; beside it is a set of spring buffers. The tube leading from the plunger downwards empties oil seepage into a receptacle. (Photo by Jim O'Boyle.)*

11

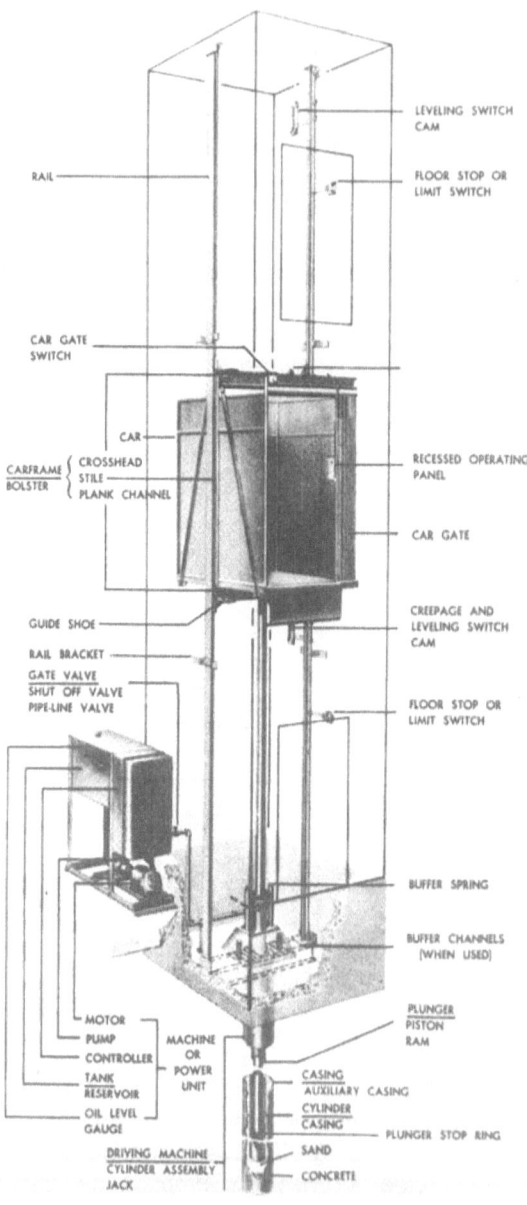

Fig. 1.2. Diagram of a typical hydraulic elevator. (Courtesy National Elevator Industry, Inc)(NEII)

12

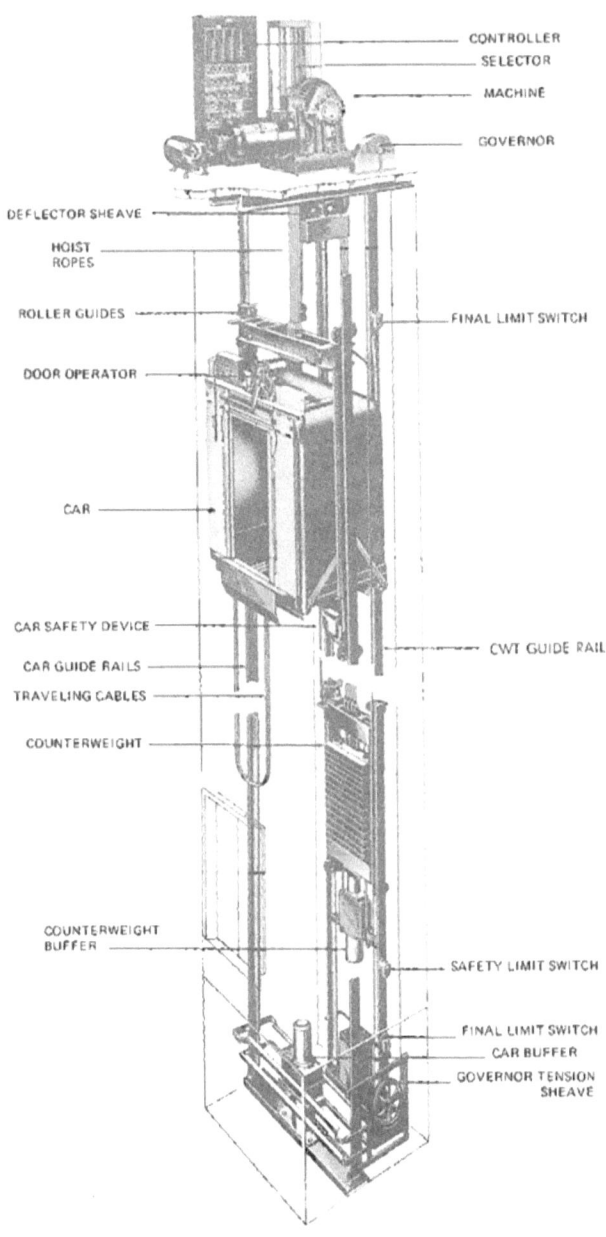

Fig. 1.3. Diagram of a typical electric traction elevator with an overhead machine room. (Courtesy NEII.)

13

Electric traction elevators operate by means of power supplied by electric drive motors. Basically, such installations consist of an elevator car, a counterweight, and a traction sheave. (See Figure 1.3.) Wire ropes -also called hoisting or suspension ropes -that are attached to the top of the car frame raise and lower the car as they pass from the car, up and over a traction sheave (a grooved wheel that guides the rope), then back to the counterweight. (See Figure 1.4.) The downward pull of the elevator car and its counterweight provide traction between the turning sheave and the hoisting ropes. The counterweights travel up and down in the hoistway on their own guide rails, moving in the opposite direction of the car as shown

Fig. 1.4. Sheaves are the grooved wheels that guide the hoisting ropes which raise and lower elevator cars. Several different types of sheaves exist; the one pictured below is the deflector sheave of an elevator with a basement machine room.

14

in Figure 1.5. Counterweights are always found in electric traction elevators, as these elevators rise several stories in a building and need to be counter-balanced. Hydraulic elevator installations traveling more than six stories may contain counterweights, while installations servicing less than six stories are usually not so equipped.

A third kind of elevator is the winding drum type. (See Figure 1.6.) This installation is nearly obsolete; however, it may still be in use in older and historical buildings. These older elevators operate by means of a spirally grooved drum onto which hoisting ropes wind as the drum turns, thus lifting the elevator car. Passenger safety has been jeopardized in these installations when safety switches failed on a turning drum.

Fig. 1.5. Counterweights are found on all electric traction elevators and some hydraulic installations_ (Photo by Jim O'Boyle.)

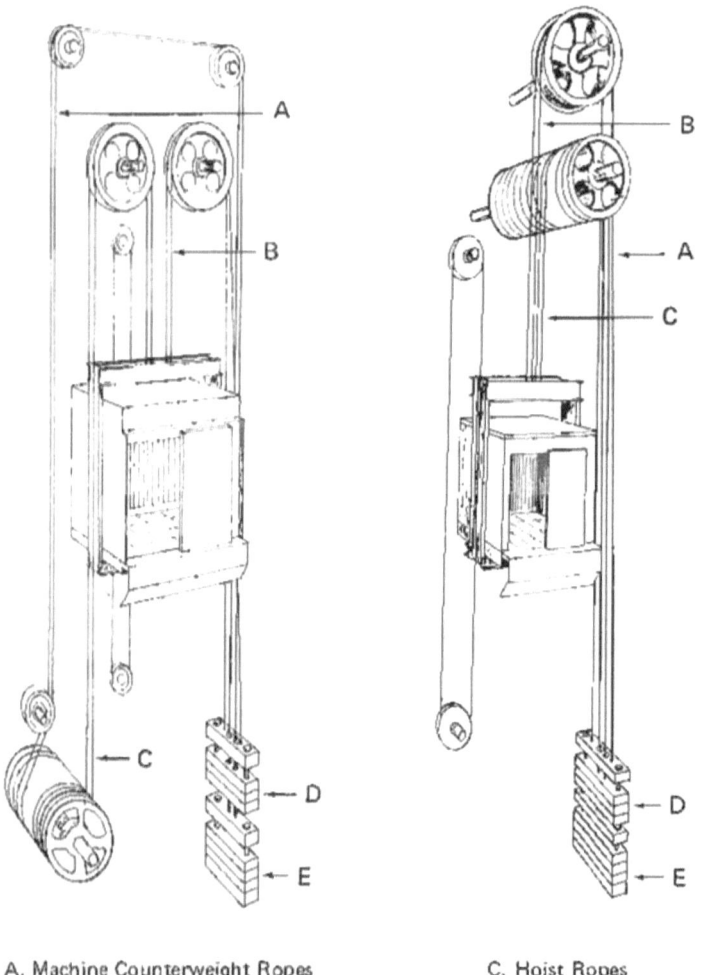

A. Machine Counterweight Ropes   C. Hoist Ropes
B. Car Counterweight Ropes   D. Car Counterweights
E. Drum Counterweights

Fig. 1.6. To the left is a winding drum elevator with a basement machine room and to the right is a winding drum elevator with an overhead machine room. Such installations are rare, but may exist in some older and historical buildings. (Courtesy NEII.)

Fig. 1.7. A *typical machine* room *in an installation with electric traction elevators (Photo courtesy of Elevator Bobs Web Site).*

The Machine Room

The machinery used to operate any type of elevator system is found in an area of a building called the machine room. (See Figure 1.7.) The machine room for electric traction elevators is usually located on the floor above the highest floor serviced by the elevator (in some cases the roof). In multibank elevatored buildings there are several machine rooms. They are usually located on the floor above the highest floor serviced by a bank of elevators, so that there are several floors with machine rooms. In some installations, the machine room for electric traction elevators is often located in the basement or on the lower floors of a building. The components of a machine room for an electric traction elevator system include a main line disconnect, a traction hoisting machine, a controller, a motor generator and a speed governor. These are the main components of an electric traction elevator machine room.

Machine rooms for hydraulic installations are usually located in the lowest level of a building, such as the basement, and generally they may be found beside the lowest elevator entrance. In some cases, however, they may be up to 50 to 100 feet from the hoistway. The components of a machine room for an hydraulic elevator include an hydraulic pump, electric motor, and reservoir, usually found in one unit commonly called a power unit, and a main disconnect and controller. Controllers may or may not be separated from the other equipment. Car safeties are usually not found on hydraulic elevators because a car will descend no faster than the hydraulic fluid can be forced out of the cylinder. Buffers usually provide adequate protection for hydraulic installations.

For both electric traction and hydraulic elevator systems, an understanding of how each of the components in the machine room functions in relation to the overall operation of the system is important to firefighters. For example, the electric power to a stalled elevator should always be cut off before hoistway doors are opened and passengers are evacuated. Firefighters need to know that this is accomplished by means of the main disconnect.

Main Line Disconnect: The main line disconnect for any elevator - usually a fused knife switch or a large circuit breaker -is generally found inside the machine room near the entrance door. (See Figure 1.8.) Operation of the main line disconnect stops the car and removes

Fig. 1.8
Throwing the main disconnect switch will remove all operating power from an elevator and stop it immediately. Shown are two common types of disconnects

all operating power from the elevator. Ventilation and lighting circuitry are usually separate from the main line disconnect, and car lights normally continue to burn and fans or blowers continue to run when the main circuit is shut off, thus assuring some measure of comfort for

18

trapped passengers awaiting rescue. (Power failure and auxiliary power systems are discussed in further detail later in this chapter.)

Hoisting Machine: Electric traction elevator cars are raised and lowered in hoistway by means of traction hoisting machines. (See Figure 1.9.) These machines operate by the interaction of a traction sheave, a driving motor, and motor brakes. Traction sheave are the large, grooved wheels that move the hoisting ropes to raise and lower traction elevators; they are regulated as to size, and must be at least 40 times the diameter of the rope. Adhesive friction (traction) enables the turning sheave to move the hoisting ropes. Traction sheaves connect to the shaft of the driving motor either directly or through reduction gears. The driving motor is a large electric motor powered by either alternating or direct current ranging upwards to about 500 volts. Thus, in the event of fire, it is necessary to observe the same precautions that apply to any fire around electrical equipment and, whenever possible, to remove the power first. Motor brakes are discussed in the section titled "Stopping Systems."

Today the gearless electronic AC hoisting machines are replacing DC units because they conserve power by reducing initial starting currents, improving train-drive efficiencies and eliminate carbon brushes. (See Figure 1.10) Adhesive friction/traction is still used with these machines to raise the car, but flat polyurethane-coated steel belts have replaced the heavy woven steel cables have been the industry standard since the 1800s. The belts are about 30 mm wide (1 inch) and only 3 mm (0.1 inch) thick, yet they are as strong as woven steel cables while being far more durable and flexible. The thinness of the belts makes for a smaller winding sheave and reducing the space required for the machine, thus making it possible to locate the machine in the hoistway. (See Machine Room-Less Elevators later in this chapter)

.

Fig. 1.9. The hoisting machine uses cables to raise and lower the elevator cars of electric traction installations. (Photo by Jim O'Boyle)

Fig. 1.10 Machine Room-Less machine being serviced by a elevator mechanic (Photo courtesy of Schindler Elevator)

Fig. 1.11. These are two types of controllers. The one on the left is an older (Photo by Jim O'Boyle.) analog unit and the one on the right is a digitally-controlled microprocessor type. (Photo courtesy of City Elevator Co.)

The hoisting machine for the hydraulic installation is the unit that contains the hydraulic pump, electric motor, and reservoir. Operation of this unit either provides oil in order to raise the plunger, or reduces the supply of oil in order to lower it. In the past, water was used as the operating fluid in hydraulic installations. Today, however, water is seldom used and oil fills most hydraulic systems. If maintenance is poor and oil leakage results, an oil puddle will form on machine room floors and in the bottoms of hoistway pits. Although most oils used in hydraulic elevator systems have high flash points, any combustible oil that puddles can be a source of fire -especially when such puddles contain accumulations of debris.

Controller: An elevator controller is the "brain" of the elevator system; it receives all signals and dispatches elevator cars in answer to these signals. (See Figure 1.11.) Elevator cars move to and stop at landings according to programming from the controller. When several elevators are grouped together, a group

Fig. 1.12. *Motor generators provide direct* current for *electric traction elevators. (Photo by Jim O'Boyle.)*

Fig. 1.13. *Shown here is one* of *the many different kinds* of *speed governors. (Photo by Jim O'Boyle.)*

controller or relay controller is most often used to coordinate the movement of all cars with selector equipment. This controller receives all calls to the elevators and responds to the calls by dispatching cars in the least possible amount of time. Basically, selectors start, stop, open, and close elevator doors at designated floors. On older elevator systems, stops, or floor bars, are arranged up and down or across a selector, and a moving pawl engages the contacts as it travels past the floor bars transmitting signals. The elevator bar and the pawl are mechanically or electrically coordinated.

Traditionally controllers and selectors were analog, today they are digitally-controlled microprocessors or computers that can be small in comparison to there older counterparts. Some controllers may have digital displays or even screens that tell you about car location. Remote elevator monitoring is becoming commonplace and newer systems may even be connected to Web based technology that during an emergency can provide vital information about the problem effecting the car operation. Generally building maintenance or managers can provide details or access to the information these systems are capable of delivering.

Motor Generator: The motor generator converts a building's alternating current to direct current. Many electric traction elevator installations use direct current (which can be controlled better than alternating current), and each electric traction hoisting machine has an accompanying motor generator. (See Figure 1.12.) Motor generator voltages are the same as those for the driving motor and, similarly, can vary up to 500 volts. As with the driving motor, water should not be used indiscriminately in the event of fire. *Also,* the same precautions that apply to any fire around electrical equipment should be observed and, whenever possible, the power should be removed first. In some installations built after the mid 1970's the motor generator may have been replaced by a silicone rectifier that performs the same function. Newer units are generally alternating current and do not require a motor generator or silicone rectifier.

The Speed Governor: All electric traction elevator systems have speed governors; during a malfunction, speed governors activate elevator safeties which safely stop the car. (See Figure 1.13.) When the machine room is above the hoistway, speed governors are located in the machine room; otherwise, they are inside the hoistway overhead. These devices read the speed of an elevator car by means of an endless wire rope (called a governor rope) which passes from the speed governor to the operating lever of the elevator safeties, around a tension sheave in the elevator pit, and back over the speed

23

governor sheave. The several functions of the speed governor are initiated by centrifugal force: i.e., at a predetermined speed, the spinning governor activates overspeed and speed-reducing switches which in turn slow down and stop a speeding elevator. Should a speeding car continue to accelerate after all switches have acted, the governor sets a gripping jaw, and the gripping jaw seizes the governor rope. As the elevator continues downward, the tensioned governor rope trips the car safeties and safely stops the car. (See "Stopping Systems" for a more detailed description of car safeties.)

Stopping Systems

Both hydraulic elevators and electric traction elevators are equipped with mechanisms for stopping the car either during normal operation or during a malfunction in the system. Such malfunctions may include a slackening in the ropes or excessive car speed. The hydraulic elevator has no actual braking system; instead, the car is brought to rest by decreasing and stopping the flow of oil into or out of the ram. In place of brakes, hydraulic systems have anti-creep devices that help keep the car stationary. Because it is difficult to keep hydraulic valves and stuffing *boxes* free from leaks, cars can drift away from landings as the oil seeps out. When the car drifts a predetermined distance, it trips a switch in the shaft, thus activating the motor, which pumps oil into the system to bring the car back to the landing. (This does not happen if the power is cut off.)

Electric traction elevators contain a braking system that is part of the drive machine. During normal operation, the brakes hold the car in place at the landing by mechanical actuation similar to the operation of brake shoes against the brake drums in an automobile. (See Figure 1.14.) During normal operation the brake is electrically released. The "emergency brake," that prevent an elevator from overspeeding even if (however unlikely) all the hoisting ropes should break are called safeties. Safeties are located at both ends of the safety plank, and they operate simultaneously on each guide rail. (See Figure 1.15.)

In addition to these methods for bringing elevators to a stop, both hydraulic and electric traction elevator installations contain final stopping devices in the hoistway pit or at the top of the hoistway. Should the car travel beyond the lowest or highest landing, these terminal stopping devices (called normal terminal and final terminal stopping devices) are used to shut off the power to the elevator. The final terminal stopping device, which is activated by the moving car, Cuts off the power to the driving motor and brake coil should the

24

controller fail to stop a car normally. As with many other safety circuits the elevator will remain inoperable until an elevator service mechanic resets the circuit; this discontinued operation of a malfunctioning elevator forces correction of the problem. During a fire, firefighters should remember that heat can affect terminal switches, as can water funning into hoistway switches and wiring. Switches and wiring affected in this way can stall elevators, trapping firefighters.

Fig. 1.14. Electric traction elevators have brakes that work something like those on most automobiles. In this picture the pointers are resting on the brake pads, which grip the drum to stop the car.

Fig. 1.15, Safeties are an emergency stopping device. When activated by the speed governor, the large spring stops the car by wedging metal against the guide rails. (Photo by Jim O'Boyle.)

25

The rated load is the amount of weight that an elevator car is designed to lift at a rated speed. The rated speed -measured in feet per minute -is the speed at which the car is designed to operate with the rated load. Rated load and rated speed vary, depending on the type of system and how high an elevator travels. The rated speeds at which some cars operate can exceed 1,000 feet per minute. Speeds of this rate are found in electric traction elevators; hydraulic elevators are not generally used for lifts of over 65 feet or speeds of over 200 feet per minute.

Fig. 1.16. *Shown here are* a-*spring buffer* (above) *and an oil buffer (right). (Photos by Jim O'Boyle.)*

Machine Room-Less Elevators

Thanks to microprocessors and the development of what are termed permanent magnet motors (PMM) manufactures are now able to produce Machine Room-Less (MRL) elevators. This technology has been utilized outside the USA for at least 15 years and is becoming the standard product for low to low-mid rise buildings not exceeding 300 feet in height where speeds of 200 and 350 feet per minute are required. MRL elevators has most, if not all, of the same components

26

as electric traction elevators except for the motor generator that is not needed with a PMM.

The arrangement of the MRL will very from manufacture to manufacture (See Figure 1.17). The motor themselves may be located at the top of the hoistway, on the side of the hoistway, or on top of the car. The controller generally is located adjacent to the hoistway, but may be remotely placed depending on the space available in the building.

These systems will become more and more prevalent in most jurisdictions as US codes adapt to this relatively new technology. ASME A17.1 code first addressed MRL's in the 2005 version and with the adoption of ANSI/ASME A17.1 the MRL is here to stay. The benefits to this product are as follows:

- Flexible space-saving configuration – No need for a Machine Room.

- Rapid installation - eliminates the cost and environmental concerns associated with a buried hydraulic cylinder filled with hydraulic oil.

- Environment-friendly - the MRL elevator will save a significant amount of energy (estimated at 70-80%) as compared to hydraulic elevators which they often replace.

- Smooth, quiet performance – The PMM traction type machine results in superior performance and ride quality compared with hydraulic elevators.

Concerns with the MRL Elevator

The largest concerns with the MRL technology for emergency personnel are elimination of the traditional ways of locating the car and controlling power during an entrapment incident. Web based system may help with locating the car and power controls for these systems my be with the controller or may be a part of the building mechanical room where the main power breakers are located.

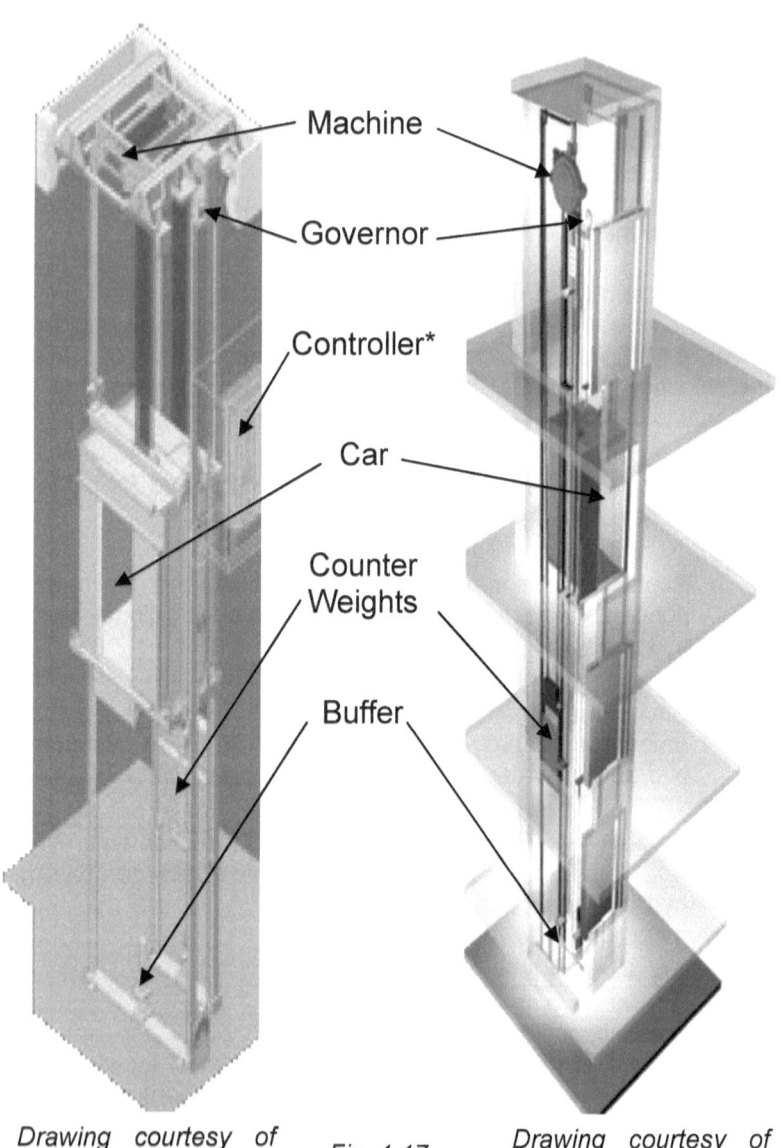

Machine

Governor

Controller*

Car

Counter
Weights

Buffer

Drawing courtesy of Otis Elevator. *Note controller shown next to hoistway but can be remote located.

Fig. 1.17
Machine
Room-Less

Drawing courtesy of Kone Elevator. *Note controller not shown can be remote located.

The Hoistway

As stated earlier in this chapter, hoistways, the vertical area of elevator travel, can be divided into three general types: (1) single hoistways, (2) multiple hoistways, and (3) blind hoistways. Unlike the earlier open shafts, most modern hoistways (other than hoistways constructed of transparent materials for observation elevators) are completely enclosed to help reduce vertical fire spread in a building. In addition, today's hoistway walls are constructed of fire resistive materials, and hoistways themselves are protected by fire-rated doors.

Single hoistways are usually found in smaller buildings, although larger office buildings sometimes have isolated freight elevators and single shuttle elevators. Often, both types operate in a single hoistway. However, most hoistways are multiple hoistways that accommodate more than one elevator. Four cars is the limit permitted in one multiple hoistway by ANSI/ASME A17.1.

Blind hoistways that serve only the upper stories of high-rise buildings do not have hoistway doors in the blind portion of the hoistways -that area between the main lobby and the lowest floor of the area served by the express elevator. In single blind hoistways, access doors should be provided, and should be no more than 36 feet apart. In any case, firefighters who become trapped in blind hoistways often find it difficult to escape.

Hoistway Construction: Today's modern hoistways are constructed of noncombustible materials such as concrete block, reinforced concrete, brick, gypsum, terra cotta, or a combination of these materials. Hoistway doors, which complete the enclosure of a shaft, help protect the landing openings should a fire occur. They also keep building occupants from falling into an open hoistway. The elevators and counterweights are guided in the hoistway by T-shaped steel guide rails; these are fastened to the building structure. Firefighters should be aware of the types of construction materials used in the hoistways of buildings in their communities because some building materials (such as reinforced concrete) are virtually impossible to breach with the usual forcible entry tools carried by firefighters.

As previously described in the "Stopping Systems" section of this chapter, there are safety devices called the normal terminal and final terminal stopping devices located at the top and bottom of hoistways. These devices, which are independent of normal controlling mechanisms, cause an elevator car to stop before it reaches its upper and lower limits of travel.

Buffers, which back up normal and terminal stopping devices, arc located at the bottom of hoistways in the pit. Should a normal or final terminal stopping device fail to stop an elevator car moving below a lower landing, the car will stop on a buffer.

Hoistway Doors

Elevators have two sets of doors through which passengers enter and leave: the outer door(s) to the hoistways and the door(s) or gate to the elevator car itself. A motor supplies the power that opens the elevator car doors when the car, scheduled to stop, enters the landing zone. The door motor, which powers only the car door or gate, is located on top of the elevator cab or car, (See Figure 1.18). As the car door or gate opens, it in turn drives the hoistway door open by means of contact with rollers on the hoistway door and a driving vane or clutch on the car door. Hoistway bi-parting doors for freight elevators are powered directly and each door has a motor mounted in the hoistway. Modern hoistway doors are equipped with electro-mechanical devices called interlocks, which lock the hoistway door(s); an elevator car will not operate unless the hoistway doors have been locked by the interlock. Interlocks are generally placed on the header beam over the hoistway opening, although some are located on the side of the opening. Upon activation of the power door motor, the clutch picks up the interlock roller, releases the interlock, and opens the hoistway doors. Reverse action of the power door motor closes both the hoistway doors and the interlock.

Fig. 1.18. This belt-driven power door operator is mounted on top of the car and opens the door at each landing. (Photo by Jim O'Boyle.)

Hoistway doors serve other useful safety purposes. For example, because they cover hoistway openings at each landing, they keep building occupants from falling into open hoistways. Hoistway doors also play an important role in fire protection. They are usually constructed of heavy metal with the door hangers and hardware generally carrying a 11/2-hour fire rating. When closed, the doors help prevent smoke and fire from entering the hoistway. However the intense heat from a fire can warp hoistway doors and prevent them from closing. In such an instance, the hoistway interlocks cannot close, preventing the elevator car from moving. Fire and smoke can then enter the elevator car and endanger the trapped passengers.

Another unsafe condition can result when locked hoistway doors fail to operate properly. In such a case, hoistway doors may sometimes be released by a special key which opens from the landing side of the doors. Unfortunately, however, most hoistway doors do not have such key-operated devices. In some states, all doors in single hoistways are required to have the unlocking device, but in multiple hoistways only the bottom landing doors are required to have such a device in order to permit entrance to elevator pits that do not have access doors. In any case, manually operated unlocking keys for hoistway doors are not always readily available; therefore, it is necessary for emergency rescue personnel to be familiar with the operation of the type of hoistway door involved. Such knowledge will enable them to force the door in the best possible manner and with the least possible damage. Modern passenger elevator systems have five types of hoistway doors: (1) center-opening, (2) two-speed, (3) two-speed center opening, (4) single-slide, and (5) swing. (See Figure 1.19.) Another type -the vertical biparting door is sometimes found on freight elevator installations.

Center-opening Door: Center-opening doors consist of two panels that move away from each other when they open, and towards each other -until they join -when they close. Both panels operate simultaneously. Locks for these doors usually can be easily located at the top of the door at a keyway near the point where the two panels meet. All locks in anyone installation will probably be in the same location.

Two-speed Door: A two-speed door consists of two panels, with one of the panels sliding behind the other. Both panels move horizontally in the same direction when opening, and both reach the open position simultaneously. The rear panel, called the high-speed panel, travels faster and farther than the front panel; it is the panel that contacts the door jamb, or upright piece forming the side of the door opening, when the door doses. The lock for the door is located at the top of the rear panel on the jamb side.

Two-speed Center-opening Door: A two-speed center-opening door consists of four panels with two of the panels sliding in each direction as described for center-opening doors and two-speed doors. Locks for these doors can usually be found at the top of the doors at the point where the two center panels meet.

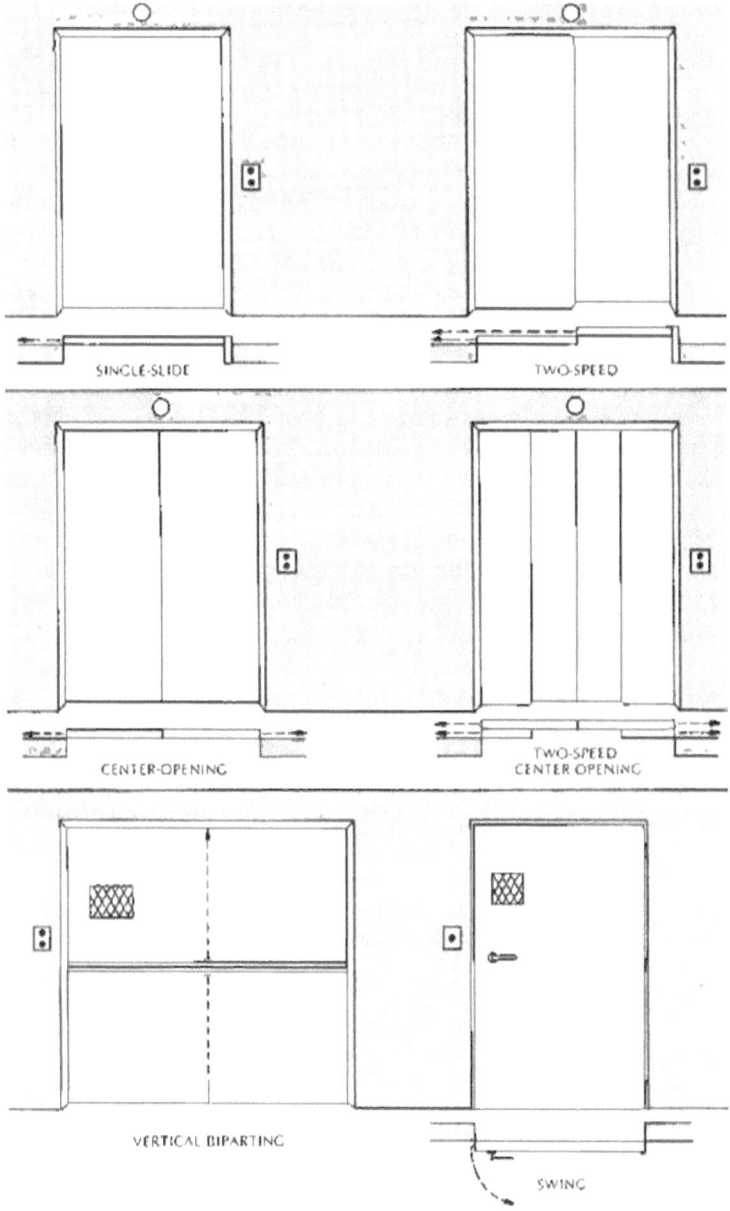

Fig. 1.19. *The various kinds* of *elevator doors.*

Single-slide Door: A single-slide door is a single panel that slides horizontally to one side of the door opening: This type of door generally operates more slowly than other types of doors. Single-slide doors are usually found in small office buildings and apartment buildings. Locks for this kind of door are usually found at the top of the door at a keyway near the point where the door panel meets the strike jamb. Locks for these elevators will probably be in the same location on all cars in anyone installation.

Swing Door: A swing door is usually found in private residences, older, low-rise buildings, some apartment buildings, and housing projects. Swing doors are doors that open outward from the hoistway. Small wire glass vision panels, usually found on the door, provide a view so that passersby will know when the elevator arrives and won't be harmed by the door's outward swing. The locks of these doors are usually in a keyway near the top of the handle side of the door, and are likely to be in the same location on all elevators in anyone installation.

Vertical Biparting Door: Vertical biparting doors, common to freight elevators, consist of two or four steel panels that operate vertically: to open, the top panel(s) moves upward and the bottom panel(s) moves downward. This movement is reversed when the doors close. Both movements occur simultaneously and can be performed manually or automatically. Locks for this type of door are usually located on one side in the area where the sections meet. If the hoistway door unlocking keyway or device is located on one door, the lock will usually be found in the same spot on all doors in that installation.

The Elevator Car

The elevator car, also called the cab, provides passenger and freight transportation to the various floors of multi-story buildings. The elevator car is usually the only means of transportation available for fire service personnel during most building emergencies. However, because power failure or a system malfunction can trap passengers inside an elevator car, the car itself can also be the site of an emergency. For this reason, elevator cars are provided with emergency exits for use during passenger rescue operations.

Construction and Ventilation: Elevator cars are assembled in the hoistway. The car's frame (or sling), in which the car rests, is

34

suspended from hoisting ropes for electric traction elevators and usually is bolted to the top of the piston for hydraulic elevators. The bottom of the frame is formed by the plank (safety plank for electric elevators), which is joined to the crosshead, or top beam of the frame, by uprights known as stiles. The elevator car's platform, or floor, rests on the plank, and is generally constructed of steel or wood with a steel frame. The exterior walls and the top of the car are made of steel or wood. When wood is used, sheet metal or fire retardant paint covers the car to act as a flame guard.

The interiors of passenger cars may contain a variety of materials intended to give the occupants a pleasant atmosphere during their ride. Natural metallic finishes, including extruded aluminum, stainless steel, bronze, and nickel-silver are used, and interior panels of wood combined with a veneer of high-pressure plastic laminate facing are popular.[4]

Elevator cars are ventilated either by natural means or by blowers. The construction and ventilation of elevator cars can have a great effect upon the safety of elevator passengers if a fire develops in a building. Elevator cars that contain combustible materials can present hazardous car conditions should a fire extend to the car. In addition, a pressure blower ventilation device can force smoke and gases from a hoistway fire into the car, thus seriously endangering passengers.

Emergency Exits: The most familiar form of entrance to and exit from an elevator car is, of course, the front entrance door or doors to the car. Since the doors to an elevator car can malfunction during an emergency situation, and often cannot be used as an exit when cars are stalled between floors, other means of egress are necessary. Thus, all elevator cars contain some form of emergency exit. One of the most common is the top escape hatch that opens outward. Top escape hatches can be useful to emergency service personnel in many emergencies. For example, they can be used in emergency rescue situations. These exits usually have an electrical contact which shuts down power to the elevator machine.

Elevators operating in multiple hoistways generally have an emergency side exit in addition to the top exit. A side exit allows for escape from, or entrance to, another car in the multiple hoistway but only after the car is aligned with an adjacent car proper safety

---

[4] Adler, Rodney, R. *Vertical Transportation for Buildings,* American Elsevier Publishing Company, Inc., NY, 1970, pp. 173-175.

precautions taken, and main disconnects for both cars opened. The emergency side exits of some cars may be panels that can only be opened by a key or a door handle. Modern elevator cars are equipped with electric contacts intended to hold the car stationary when an emergency side exit is being used; however, to be completely certain the car will not move, the power should be shut off in the machine room prior to using the side exit.

Car Doors

With the exception of swing doors, elevator car doors are most often the same type as the hoistway doors: i.e., a center-opening hoistway door will operate with a center-opening car door, while a single-slide hoistway door will operate with a single-slide car door. Swing doors, however, will either have single-slide car doors or manually or automatically operated gates.

Car doors are generally made of metal and do not have fire protection ratings. These factors make passenger safety dependent upon the fire resistance of the hoistway and hoistway doors should a fire Occur near the elevator. A hoistway door that has jammed as a result of heat from a fire may seriously endanger the occupants of a car. Once fire enters a hoistway, both the elevator car and its doors offer little protection to the car's occupants.

Automatic Closing of Doors: Elevator car doors open and close automatically when an elevator car, programmed to stop, enters a landing zone or comes to a stop at a landing. The landing zone is an area extending from a point 18 inches below a landing to 18 inches above a landing. As previously explained, a door motor located on top of the car supplies the power for opening and closing the car doors on passenger elevators. Although car doors usually do not have locks, they must be shut and an electrical contact made before an elevator will operate.

Door Reversal Devices: Door reversal devices are used to protect entering and exiting passengers. These devices operate by contact with a person or object, by a light source or photocell, or by an electronic sensing device. The safety edge mounted on car doors to keep passengers from being bruised may be supplemented by a light source and photocell mounted on opposite sides of the car entrance: a passenger walking through the doors as they are closing interrupts the light beam, which automatically reverses and opens the doors. The electronic sensing device works by means of a balanced electrical circuit: i.e., when the balance of energy is upset by an intervening object or person, there is a break or drop in power that triggers the doors' reopening.

Electronic detectors can also be used for door reversal. New systems may have timing devices which, after a predetermined period of time, sound an audible signal and close the door at slow speed, overriding any reopening mechanism(s).

Door Restrictors: Door restrictors are a device that can cause problems during fires and rescue procedures. While they serve a useful purpose to keep the passengers safe by not allowing them to open the car doors should the car stall in the hoistway and not be close to a floor landing. They can and do hinder attempts to remove trapped passengers as they either mechanically or electro-mechanically lock the interior car door in the closed position.

Most all door restrictors can be defeated provided firefighters know the type of system and how to deactivate its operation. Generally this has to be done on the top of the can once the exterior doors are opened or access is gained to the top of the car.

A note from 9-11 -- USA TODAY found:[5]

- **Newly installed safety devices condemned many people to death.**

    To comply with building codes, the World Trade Center since 1996 had been adding locks that made it impossible for passengers to force open the doors of stalled elevators. These locks, called "door restrictors," had been added to about half of the 198 elevators in the twin towers. Nobody is known to have escaped from an elevator locked by a door restrictor. The World Trade Center followed a long-established approach to elevator rescues: Leave people inside stalled elevators until professionals can perform rescues. The elevators had three mechanisms, including the restrictors, designed to prevent people from accidentally falling down elevator shafts. An untold number were still trapped when the buildings collapsed.

_____

[5] USA TODAY 9-4-2002

Fig. 1.20 Collapsible Door Restrictor works in conjunction with a hoistway door angle to deter passengers from exiting the car outside the landing zone. It is designed to replace existing car angles to allow easy access to the car top to perform maintenance or evacuate passengers (Photo courtesy of Otis Elevator)

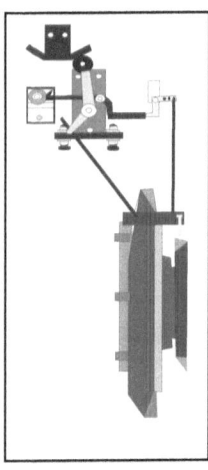

Fig. 1.21 Mechanical Door Restrictor - This door restrictor mounts directly on the elevator car door. Rescue personnel can disengage the restrictor when needed to evacuate trapped passengers

Fig. 2.22 Electro-Mechanical Door Restrictor: There are a variety of these types of systems most of which can be deactivated by disconnection of the power to the system. Some are battery operated which makes this job simple. Some systems are deactivated upon the use of firefighter or independent service. (Photo courtesy of Otis Elevator)

THE ELEVATOR CONTROLS

The elevator door and the elevator car are monitored or activated by the car and hallway fixtures. The most common hallway or landing fixture is the push-button station. Used to call a car, it may also indicate direction of movement. An additional hallway fixture, a position indicator, may be found in the elevator lobby. There may be a similar device inside the elevator car. These position indicators are valuable to fire service personnel for the following reasons: (1) they can show the general location of a stalled elevator, (2) they can be used to monitor the car location from the inside when the elevator is being used to transport firefighters during a fire, and (3) they can be used to indicate to the officer at the lobby traffic direction station or fire command post, the location of firefighters enroute to or from the floor of firefighting operations.

The next most common interior car fixture is the car panel with its selection of signals. The car panel can be used to select a desired floor, to show direction of travel, and to close and open the *doors*. Car signals are also available for emergency use by passengers and rescuers. Passenger emergency signals (stop switch and alarm) are located on the car panel and are operated manually (See Figure 1.23.). Fire service personnel emergency controls, located in both the car and in the lobby, are key-activated.

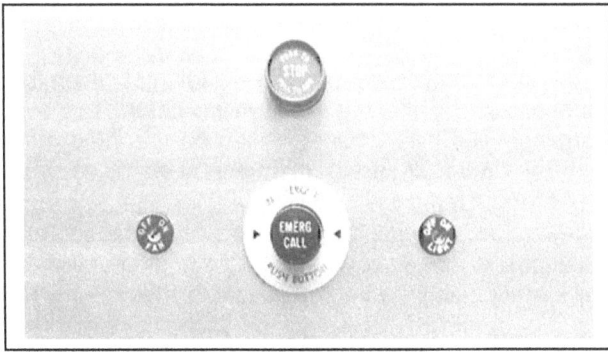

Fig. 1.23. The Alarm and Emergency Stop switches appear on all elevators.

In addition to emergency signaling systems, a means of verbal communication is provided in most modern elevator cars. When elevator cars were operated manually, they often contained telephones for communication between the car operator and the building personnel. Conventional telephones are still kept in some automatic elevators for passenger use, but many modern elevators contain a form of two-way communication such as in intercom system. Loudspeakers in the cars are connected to equipment at the starter's station, building engineer's office, or some other convenient location. The starter or other authorized personnel can listen when called and speak to passengers through this equipment.

What ever its form, a communication system is valuable to both passengers and emergency personnel who must communicate with trapped passengers. Without some form of communication, trapped passengers might attempt to leave a stalled elevator car and place themselves in extreme danger. A communication system can help prevent such situations when used to convince trapped passengers in stalled elevator cars not to leave the car, to alert passengers when the operation of their elevator car has been taken over by the fire department from the main lobby thus overriding their selections, and to calm passengers and assure them that help is on the way.

Non-emergency Signaling Systems: Passengers obtain elevator cars for use by "calling" them with hall call buttons located at elevator landings. Most intermediate landings have two call buttons -an Up button and a DOWN button -which passengers use to summon an elevator car. 'Some systems have only one call button, as do the top and bottom landings. Inside the elevator car, passengers register calls on buttons numbered or lettered to correspond to the floors serviced by the elevator. Some types of buttons illuminate to give visual indication that calls are registered; as the elevator answers each call, the light goes out.

Inside the car, the direction of movement and the landing at which the car has stopped are shown by the car position indicator, if one is provided. At the landings, a similar indicator may be mounted over the entrance of each car and/or separate directional lanterns may be used. Directional lanterns may be located above the elevator entrance, on the side jamb of the elevator entrance, or on the corridor wall. In some installations, a simultaneous audible signal is activated to attract passenger attention as the directional light flashes on.

Emergency Signaling Systems: Passengers in a car that has stopped suddenly or that operates improperly can alert building service personnel to these conditions by activating an emergency alarm button on the car control panel. Printed near or on this button are such statements as: "In emergency push button," "Alarm," "Emergency Alarm/Call," "Emergency," or other similar messages. In most cases, the button is red. Activation of this emergency alarm button usually sets off an alarm signal that notifies building personnel and building occupants of an elevator malfunction.

Another type of control device is the "Firefighters Service" control. This control, located on or beside the car's control panel, is key activated by turning it to the On position, after turning the main floor Lobby Switch of the Firefighters Service to On also. Activation of the system places the car under the control of fire service personnel traveling in the car, once the Lobby Switch has been activated. This switch is a vital life-saving device during fire and smoky conditions. Should a car door not be on Firefighters Service and its doors open at a floor landing that is on fire, the smoke could activate the photocell and so prevent the doors from closing. Activating the Firefighters Service will override all door closing devices. (For a fuller treatment of this feature, see Chapter 4.)

In some states, fire service personnel and other rescue personnel can control car operation in an emergency by manual operation of an independent service system. The controls for this independent service system are kept locked. When the switch for this type of service is located inside the car, it is usually contained in a locked panel. This switch may be identified by the initials 1.S. (Independent Service), or the labels or signs "Manual," or "Hospital/Medical Service." In installations where the I.S. switch is not located in a locked panel, the switch itself will require a key for operation.

What ever the type of elevator installation, hydraulic or electric traction, elevator cars, hoistways, hoistway doors, car doors, and elevator controls are essentially the same. And although the operating components for each type of installation may vary, the emergencies that occur happen in either type. Every component of an elevator installation is critical to a system's safe operation.

In emergencies, it is necessary for emergency rescue personnel to have a basic understanding of the operation of elevator cars, doors, and controls. These three components are the most visible parts of the elevator system. They are also the components used most during any elevator rescue or emergency situation. While invaluable, such an understanding of the basic elevator components is only one factor in preparing to handle elevator emergencies. Emergency rescue personnel also need to be informed as to the type of elevator service used in an occupancy, and should be aware of the peak traffic flow (or occupancy use patterns) of the elevator and what kind of passengers use the elevator.

SAFETY PROVISIONS FOR ELEVATORS

Maintaining the day-by-day safe operation of a building's elevator system is usually the responsibility of the building superintendent or plant manager. Inspection service by elevator companies and elevator mechanics is also available. During such inspections, car machinery, emergency switches, car leveling ability, bumpers, and buffers, etc. as well as the general housekeeping of the entire system -are checked for their safe operation.

Many states, cities, and insurance companies inspect and test elevators on a routine schedule, The test and inspection usually follow procedures outlined in ANSI/ASME A17.1, published by A.S.M.E. From their earliest days, elevators have been designed to preclude possible hazardous situations. Safety provisions include door operation devices, emergency exits, fire protection ratings of doors, manual controls for operation in the event of an emergency, and alternate power supplies in the event of a power failure. Today, as a direct result of this concern for the safety of elevator systems, vertical transportation is safer than any other means of travel.

Overspeed

As noted earlier, an integral part of every electric traction elevator system is the speed governor which prevents an elevator car from descending too fast by means of a cable called the governor rope. The governor rope passes over the sheave of the speed governor at the top of the hoistway and under a tension sheave at the bottom of the hoistway, thus forming a continuous loop. Should an elevator begin to overspeed, the speed governor will trip a safety switch at a certain set speed. This action usually slows down and stops the car. Should the car's descending speed not decrease, the speed governor trips and clutches the governor rope with enough force to set the safeties on the car, forcing them to grip the guide rails. Further motion of the car makes the safeties wedge between the guide rail and the car until the safeties bring the car to a smooth stop. After the car stops, the safeties continue to grip the guide rails and hold the car until manually removed by an elevator mechanic.

Overtravel

Protection against overtravel is provided by normal and final terminal stopping devices, which are located near the upper and lower ends of the hoistway. A car that continues to move past the top or bottom of a landing trips the final terminal stopping device, cutting off power to the drive machine and applying the brakes on electric traction elevators. If the car continues to move, the final limit switch device will be activated, cutting off power to the drive machine and applying the brakes on electric traction elevators. When this switch is tripped, the car can be moved manually only, and only by an elevator mechanic. Should an elevator continue to descend past its lowest landing despite the activation of terminal stopping devices, the car is brought to a stop by the bumpers or buffers in the pit.

Power Failure

As a precaution against an electrical power failure in a building, emergency power systems for modern elevators assure at least a minimum of energy for elevator lighting and communication. These systems are particularly important for passenger safety since they enable fire service personnel and building personnel to reassure passengers trapped in automatic elevators stalled by power failure. Emergency power for lighting also contributes significantly to passenger reassurance. The power source for these systems is provided by a charger-storage, battery-converter combination that supplies alternating current for fluorescent or incandescent lamps and direct current for telephone or intercommunication equipment. When outside power fails, the emergency system automatically starts its converter and takes over the lighting and communication electrical load.

In some buildings, emergency power is available to operate one car at a time, at a reduced speed by either manual or automatic power. (See Figure 1.24.) Most Model Building Codes require emergency power to operate at least one elevator in high-rise buildings. Manual selection may be made by a key-operated switch in the building lobby; however, operation of the manual selector switch depends upon what personnel are available. With automatic selection, cars start automatically on emergency power, one at a time, and move to the main floor.

43

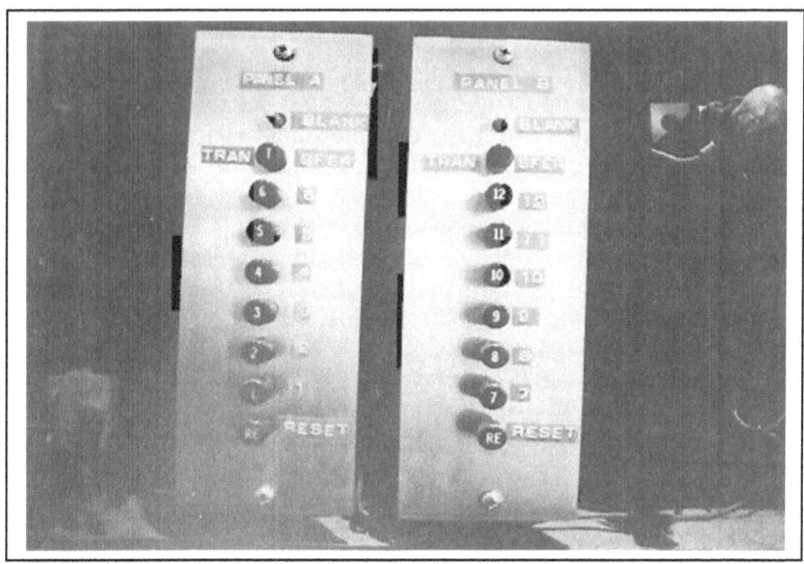

Fig. 1.24. *This* is *an example of* a *panel that operates elevators, one at* a *time, on emergency power. After* a *car is called to the ground floor, the next car is called by pushing the 'reset' button and then pushing the floor button.*

UNSAFE CONDITIONS

Despite the impressive safety record and the safety provisions of elevators, there are serious elevator emergencies that cannot be corrected by switches or by use of another power source. These emergencies include the operation of cars when there is a fire, the rescue of trapped and injured passengers, the rescue of persons pinned in hoistways, and the rescue of persons who have fallen into hoistway pits. It is well worth repeating that, whenever possible, elevator emergencies should be handled by an elevator mechanic or manufacturer's representative with the assistance of department personnel who are trained in elevator emergency procedures. There is no substitute for the training and experience which these persons possess. Their knowledge of the complexities of the elevator system ensures greater safety for passengers inside the car. Only in extreme emergencies should a rescue be undertaken without an elevator mechanic or qualified manufacturer's representative present.

CHAPTER 2

Preplanning for Emergencies

ELEVATOR SAFETY

Modern elevator systems have an outstanding safety record, especially in relation to the distances they travel and the number of individuals they transport. This record is due in part to the safety equipment that is built into these systems. However, natural disasters, improper use, and other emergencies can place elevator cars and passengers in situations in which even the best safety equipment becomes ineffective. For example, natural acts such as hurricanes, tornadoes, floods, and wind and electrical storms can result in general power failures. In one large city in the eastern United States a power failure caused thousands of elevator passengers to be trapped in cars stalled between floors. The stranded passengers had to wait several hours before they could be rescued.[6]

Stranded passengers are, however, the least severe result of an elevator malfunction or emergency. In some incidents {ire and the products of combustion have caused the death of individuals who have tried to use elevators during fire conditions. In July, 1971, for

---

[6] "Elevator Emergencies" Special Interest Bulletin No. 55, Feb. 1974, American Insurance Association, New York.

instance, a motel fire in New Orleans, Louisiana, killed five individuals who attempted to leave the motel's fifteenth floor by elevator. When the elevator they were riding reached the twelfth floor, a type of mechanical failure (which will be described later in this chapter) caused the doors to open onto the fire. All five passengers died from the heat and smoke in the twelfth-story corridor.[7] Also in New Orleans, on November 29, 1972, a fire in a high-rise, general-purpose building resulted in the deaths of elevator users. Three men, intent on assisting people on the building's upper floors, boarded an elevator. When the car reached the fifteenth floor -the fire floor -the doors opened and the three men were knocked to the car floor by the inrush of hot gases. Two of them were able to survive by lying face down on the floor until they were rescued, although one of these men died several days later. The third man, who lost consciousness when he was knocked down, fell face up on the floor. He died of smoke inhalation before he could be rescued.[8]

Not all building fires have resulted in deaths to elevator passengers. The fire in 1974 in San Paulo, Brazil is proof of this fact. Of the 422 occupants who survived the Joelma Building fire, the majority made their escape by elevator. The success of elevator use in this fire was attributed to two conditions: (1) the use of express service by elevator operators, and (2) the uninterrupted power supply to the elevators.[9] In general, elevators should not be used to escape from a building fire unless they are equipped with the Firefighters Service (discussed in Chapter 4). Even then, passengers should be evacuated using elevators only when absolutely necessary and under the direct supervision of fire personnel.

Reasons for Fatalities

During a fire, it is unlikely that a person can safely exit by using an elevator. Under fire conditions, an elevator can become a death trap for the car occupants. There are three situations in which this can occur. First, an individual may attempt to escape from the fire floor by signaling and, if the power supply has not been affected, obtaining an

---

[7] 'Watrous, Laurence D., "Fatal Hotel Fire," *Fire Journal,* Vol. 66, No.1, Jan. 1972, p.8.

[8] 'Watrous, Laurence D., "High-Rise Fire in New Orleans," *Fire Journal,* Vol. 67, No.3, May 1973, p. 9.

[9] 'Sharry, John A., ··South America Burning," *Fire Journal,* Vol. 68, No.4, July 1974, p.27.

elevator car. However, smoke and the products of combustion usually affect door closure: the photocells on elevator doors react to smoke and prevent the door from closing. The force of hot gases may also prevent door closure, and the heat generated by the fire can warp the doors or affect the wiring at the interlock thereby preventing the car from moving. In addition, the lives of passengers already traveling in the car may be endangered if a car descending from floors above the fire responds to the signal on the fire floor.

Elevators may also become death traps if they are used by persons who, unaware that a fire exists on a certain floor, respond to an emergency signal from that floor. This situation occurred in 1972 at the Baptist Towers Home for Senior Citizens in Atlanta, Georgia. At 2 A. M. a guard responded to a buzzer on the emergency call system from the seventh floor. He took the elevator to that floor. Firefighters later found the guard's body in the elevator at the seventh floor. The guard probably did not know there was a fire on that floor. [10]

Still another way in which elevators become death traps is if they are affected by the side effects of fire conditions. For example, during a fire that is being fought by firefighters, the signaling system of an elevator can become activated. This situation can be caused by faulty hall buttons which have been short-circuited by the heat from a fire, moisture from water used in fire fighting operations, or the smoke particles present during a fire. As a result, unsuspecting elevator occupants are sometimes confronted with hazardous fire conditions and doors that do not close.

In the following excerpt from *Fighting High-Rise Building Fires - Tactics and Logistics,* former NYFD Deputy Chief Robert F. Mendes describes some of the ways elevator systems can complicate the fire fighting effort and, in some situations, even contribute to the fatality rate in high-rise building fires: [11]

> The elevator system is also a problem of large proportions, for not only has it proved itself to be the occasion of many of the fatalities in high-rise fires, but its malfunctioning may cause the complete breakdown of the logistics effort upon which a successful fire fighting effort depends. In addition,

---

[10] 'Willey, A. Elwood, "Fire, Baptist Towers Housing for the Elderly," *Fire Journal.* Vol. 67, No.3, May 1973, p. 17.

[11] Mendes, Robert F., *Fighting High-Rise Building Fires -Tactics and Logistics,* National Fire Protection Association, Boston, 1975.

any breakdown of an elevator, whether it is carrying firefighters to positions of strategic need or carrying occupants to safety, acts immediately to divert large and specialized forces from fire fighting efforts to new and unanticipated rescue efforts.

Elevator shafts are long noted as one of the obvious vertical channels by which smoke and heat may travel within a building, and this feature of the system may seriously complicate what is apparently, at the outset, only a localized fire.

Freight or service elevator cars, and passenger cars when they are used to transport materials, pose an additional danger. When furnishings or supplies are moved in or out of a tenanted space, these cars contain a considerable amount of combustible material. The practice of lining these cars with quilted or padding when moving materials adds further to the fire load. This moving hazard subjects all floors of a building to danger at any time.

Much attention has been given to call buttons which are reputed to actuate the response of an elevator car by the heat of the user's finger. This is an erroneous description of the action of systems which operate with either a very slight pressure on the call button, or by the completion of an electronic circuit to ground which the user achieves by the mere touch of his finger.

However, the sudden failure of power sources, the direct attack of fire or heat on elevator system wiring or call buttons, the unknowing and unfortunate dispatching actions from the control panel in the lobby, the buckling or distortion of elevator shaft door mechanisms, the overheating of elevator machinery rooms, and, finally, the actions of occupants themselves on the fire floor and at other floors in using call buttons that sequentially require the elevator car to respond robot-like, regardless of danger, to its imposed program, are among the causes of elevator system malfunctioning which complicate the fire problem in high-rise buildings.

It should be emphasized that in modern elevator systems where Firefighters Service has been installed in accordance with the requirements outlined in ANSI/ASME A17.1, most of the problems described by Chief Mendes could not happen. The system requires smoke detectors in elevator lobbies which, when activated, return elevator cars nonstop to the main lobby where they remain for fire service personnel to operate manually, if necessary. Activation of a smoke detector places all hall call buttons out of service and deactivates stop buttons as well as door-opening devices subject to heat, flame, or smoke. Cars then go directly to the main lobby without stopping to answer any registered car or hall calls.

Elevator Use by Occupants During Emergencies

The fatalities that occur in elevators can partially be attributed to the fact that occupants in elevatored buildings are usually unfamiliar with any other means of exit from the building. Most people leave a building the way they carne in -often by way of an elevator. Plaques or stickers placed on or near the Outside of elevator doors that warn against using elevators during a fire can help reduce elevator use in fires. (See Figure 2.1.) The interiors of some elevator cars are equipped with plaques or stickers with a similar warning.

Stairways marked with exit signs, and floor plans posted in elevator lobby areas that show the fastest escape from a particular floor can be valuable aids in preventing the use of elevators during fires. Regularly scheduled building fire drills that mandate the use of stairways are also valuable in familiarizing building occupants with alternate means of escape. However, fire service personnel responding to an emergency in an elevatored building should be prepared both to find elevators in use and to rescue individuals who have been trapped inside them. Fire service personnel should also be prepared to use elevators during fires and should know how to protect themselves in an elevator during a fire.

Fig. 2.1. This is one kind of sticker used to warn against using elevators during a fire. Others are being developed which have pictographs instead of words to convey the meaning.

Elevator Use by Firefighters During Emergencies

In the past, rules have stated that elevator use during a building fire should be avoided. Today, such rules are often impractical especially for firefighters who must respond to fires on the upper floors of high-rise buildings. Although aerial ladders and elevating platforms are valuable apparatus for fighting high-rise building fires, they are not generally used for emergency evacuation of occupants. Thus, while people may be prevented from using elevators during a fire, firefighters and officers may be forced to depend upon them. They should therefore be prepared to protect themselves during an emergency, and this preparation involves preplanning and training. Preplanning includes learning the types of elevator service in a building and what procedures should be used in an emergency; training involves both constant review of elevator components, and drills in performing simulated emergency operations. By addressing itself to preplanning and training procedures, this chapter will explain how firefighters who must use elevators in emergencies may prepare themselves to better cope with such situations.

ELEVATOR EMERGENCY PREPLANNING

The importance of preplanning cannot be overemphasized. Just as the fire service must plan to handle major emergencies, fires, rescues, etc., so must it preplan for elevator emergencies. However, it is often difficult to establish a priority for new preplanning programs. Therefore, rather than develop a program that will add to an already heavy workload, elevator emergency preplanning should be integrated into existing programs such as building inspection, survey, or prefire planning programs. This may help to eliminate any resistance that new programs sometimes encounter, and still place a fairly high priority on the newly developed program. One good way to develop an elevator emergency program is to base it on examples and data from previous emergency situations in which elevators were involved.

The Preplanning Survey

Before preplanning can be attempted, it is necessary to conduct a survey of the community's elevator installations. To accomplish this, the program planner must contact either the owners or managers of buildings in the area, and the state or local elevator inspector. The latter will know which buildings have elevator installations and what

50

kind, as well as how the equipment operates. The relationship between the fire department and a building's owner, manager, and at times, its maintenance personnel, is often delicate, calling for the utmost tact on the part of fire service personnel. Although many persons in charge of buildings are receptive to constructive interest in their facility, there are others who are apprehensive about such interest. In either case, the program planner should be ready to handle questions concerning fire fighting strategy for the building and reasons for the fire department's interest. Each of these issues should be dealt with separately. The program planner should stress that the intention of any preplanning survey is to improve the ability of emergency personnel to use the elevators should they be called upon to free persons in stalled elevators or fight a fire.

Often, when the matter of removing persons in stalled elevators is discussed, building owners or managers protest that their building's maintenance personnel are trained and equipped to remove trapped passengers. Fire service personnel should reply that managers can be held liable for any injury that occurs during such rescues and explain that in order to deal effectively with any elevator emergency, it is necessary to provide a sufficient number of people with adequate training and equipment. Once this has been explained, most building managers will be more than willing to concede to the fire service the responsibility for the removal of trapped elevator passengers. While conducting the preplanning survey, the importance of obtaining adequate information should be stressed. It should also be remembered that usually the building management or owner is not obligated to provide any assistance. The fire officer making the survey is a guest in their facility.

The responsibility for developing a new program should be set down in writing and then enforced. Once developed, the fire department administration is responsible for enforcing the program.

Battalion Chiefs Responsibilities: Battalion chiefs are responsible for the coordination and administration of preplanning conducted within the battalion district. The battalion chief should maintain a current emergency preplan for elevators on each target hazard, [12] public assembly hall, school, hospital, condominium, apartment

---

[12] Target hazard is determined by size, occupancy, life hazard, or degree of tactical difficulty. Standard inspection frequency for target hazards should be four times per year.

complex, industrial and mercantile establishment, and petroleum facility within the district.

Battalion chiefs should review each preplan prior to approval, making note of special problems such as location of and access to the building. Coordination of preplans involving more than one battalion district should be the responsibility of the battalion chief having basic jurisdiction.

Company Officers' Responsibilities: Company officers should" be responsible for establishing priorities with the approval of the battalion chief. They should see that all necessary data and information are properly recorded and forwarded to the battalion chief, set up appointments for preplanning surveys, and make certain that all survey personnel are properly informed.

Training Responsibilities: Preplan forms should be available for use by fire suppression personnel who have been instructed in how to complete them. Training should include a review of the completed prefire plans to determine overall standardization. The training officer should keep a copy of all preplans on file for reference purposes. ¥/here unusual circumstances arise concerning proper methods of combating fires or handling rescue problems, proper training in procedures is necessary.

Fire Prevention Responsibilities: The fire prevention officer should serve as a technical advisor to station personnel in matters of on-site fire protection equipment and occupancy of buildings. He should be able to direct them to any pertinent information that may be available in fire prevention files. Fire prevention personnel should be familiar with all elevator preplanning reports and should notify the appropriate stations of any changes affecting preplanning efforts. The fire prevention officer should receive any written complaints of fire hazards noted during the preplanning survey.

Arranging for and Conducting the Preplanning Survey

Following are some suggested guidelines for use when arranging for and conducting preplanning surveys.

Contacting the Management: The first step in preplanning is to compile information on the elevators used in a community. This information is readily available from state or local elevator inspectors. Touring a building's elevator facilities is the best method of obtaining first-hand information. However, a fire department's initial attempt to contact the building management may meet with some resistance, and the elevator companies involved may share management's feelings. Both are concerned with liability. Despite resistance from either party, the fire department representatives should point out the advantages of firefighters having knowledge of the elevator installation in order to perform rescues and save lives during fires.

Timing the Preplanning Survey: After contacting the management of the occupancy, the next step is to set an appointment for the survey. First, explain the purpose of the visit and what will be done. Be aware of the timing for your survey: light usage periods, such as nonworking hours and weekends are recommended as the best times. Before meeting the appointment, make sure to bring enough supplies and equipment (such as clip boards and preplan forms) for use during the survey. If prefire plans have been made for the facility, bring them with you. Before leaving company quarters for preplanning, telephone the dispatcher to give them the location of the area to be inspected.

The Approach to the Property: The preplan should contain information on the approach to the property, noting the streets that can be used as the best response routes to the building.

Introduction to Management: The fire department representative should contact the management or individual with whom the appointment was made. He should again explain the purpose of the survey and request permission to make an elevator preplan tour with an elevator mechanic from the company which services that facility.

Routing the Survey: In routing the survey, try to map out a route that will cause the least amount of business interruption. The tour should be systematic, and should cover all pertinent areas such as the machine room(s), hoistway(s), lobby, and pit area(s). During the survey, request an elevator car that can be used to help develop operational procedures and to check car and hoistway construction.

Specific Points to Cover: The following areas should be considered during the survey of the facility's elevator system:

• *The Machine Room*

1. Note the type of elevator (hydraulic or electric) and the location of component parts.

2. On both electric traction and hydraulic pressure elevator installations, check to be sure that the main electrical control for each car and the components that work with each unit can be identified. These should be marked with large letters or numbers. If they are not marked, suggest to the person in charge that this distinction be provided.

3. Check to see that the telephone number of the elevator maintenance company is posted in the machine room. If it is not there, suggest that it be posted.

• *The Lobby*

1. Identify the type of features used to call the cars to the ground floor, whether the Firefighters Service is provided, the number of keys, and the exact location of key cylinders. (Are these features found on more than one floor?)

2. Note whether or not there are position indicators to locate the approximate position of the car should it stall in the hoistway.

3. Learn whether the indicators have a source of emergency power.

4. Determine whether the elevator system is tied into an emergency power source, such as a generator. &e all cars tied into the system? If the panel to the emergency power switches is secured or in a remote area, a system for gaining access should be devised.

5. Check to see if there is a Firefighters Service operational instruction sign in the lobby next to the key switch. If not, request that one be provided.

• *The Elevator Car*

1. Enter the car and check the location and type of emergency exits. Identify any special tools or equipment that may be needed to open these exists. Top escape hatches can only be opened from the top of the car.

2. Place the car into the emergency mode that would be used during fire operations.[13] Check and write down the prescribed method of gaining complete manual control of the car (to be discussed in detail in Chapter Four).

3. Note the method or methods that may be used to stop the car should it overrun a desired location. However, do not attempt to operate the car using this method unless accompanied by an elevator mechanic.

4. Note the number of keys needed for car control and where these keys are kept in the building. Learn which cars axe shut down after a certain hour and are thus unlikely to be readily available for fire department use.

5. Run several test situations to make certain the procedures are correct. Learn which, if any, floors are keyed off so as to prevent access to that level.

6. Check all other cars to make certain the same conditions apply.

• *The Doors*
In a routine or complex rescue as well as during a fire the knowledge of doors and their operation is an important part of emergency procedures.

1. Enter the car. Take the necessary steps to leave the car with outside assistance. Stop the car between floors by setting the Emergency Stop Switch.

2. Try to pull the doors closed without mechanical power. If the door don't open they are most likely equipped with door restrictors. (See Chapter 1, page 35 about door restrictors)

3. If the doors are equipped with through-the-door interlock release tools or devices, obtain these devices and work

---

[13] In areas or states where a Firefighters Elevator Key Switch is used, firefighters should be aware of its features and operation in buildings located in their district. These installations can vary with code changes, manufacturer's modifications, and elevator car or lobby panel style.

with them from the outside, being certain to open the main line disconnect first.

4. Check the type of guard reopening devices on the door edges for possible sharp edges capable of cutting.

• *The Top on the Car* (N. B.: An elevator mechanic should be present during these operations.)

1. Note the size, design, and location of top-of-car exits.

2. Note the strength of the car top -how many people could it safely hold?

• *The Pit*

1. Check the best way to obtain access to the pit. Note the location of lights and stop switches.

2. Note the condition of the pit. If necessary, suggest to the owner or occupant that it should be kept free of accumulated debris and, in the case of hydraulic units, of any puddled oil.

Elevator Information Forms

An important aspect of preplanning surveys is to add detailed information obtained from the survey to the prefire plan itself. For instance, the prefire plan should tell fire service personnel responding to an emergency in an elevatored building how to obtain elevator cars for their immediate *use*. Some fire departments use an elevator information form such as the one shown in Figure 2.2 which, when completed, combines some of the information obtained from the preplanning survey with emergency operating instructions. Other departments distribute booklets detailing step-by-step procedures for handling particular types of elevator emergencies. However elevator information is recorded, details of the system in a particular building should include the following information:

Location: The building's title, address, district number.

Type of Elevator: Whether the elevator is hydraulic, electric traction or drum-type, and the type of machinery that operates the car.

Number of Elevators: The number of elevators in a building. This section can also include elevator banks, elevator zones, and the presence of blind hoistways.

Multiple Hoistway: Whether an elevator shaft contains more than one elevator car.

Manufacturer; The name of the company that installed the elevator; i.e., ThyssenKrupp (Dover), Otis, Schindler Group (Westinghouse & Haughton), Kone (Montgomery), Mitsubishi, Hitachi, Fujitec, etc.

Service/Maintenance Company: The company that repairs or maintains the elevator. *Location of Company:* The location - particularly the address, if available -of the elevator service or maintenance company.

Emergency Telephone Number: The telephone number of the service or maintenance company. Usually the company will have a 24-hour telephone number.

Machine Room Location: The location of the machine room for the elevator: i.e., roof (penthouse), basement, etc.

Firefighters Service: Whether or not the elevator has provision for Firefighters Service. Whether there is an operational instruction plaque in the main lobby.

Location of Firefighters Service: The location of the Firefighters service key cylinder for calling elevator cars to the main lobby.

Control Keys: The number and location of control keys necessary for manual operation of the car. Sometimes as many as three or four may be needed to operate the emergency service.

Interlock Release Tool: The location of the interlock release tool(s). Note whether all elevators use the same type of tool. (Note:

Interlock release tools and elevator keys should not be located where the general public would have access to them. This includes their location behind glass.)

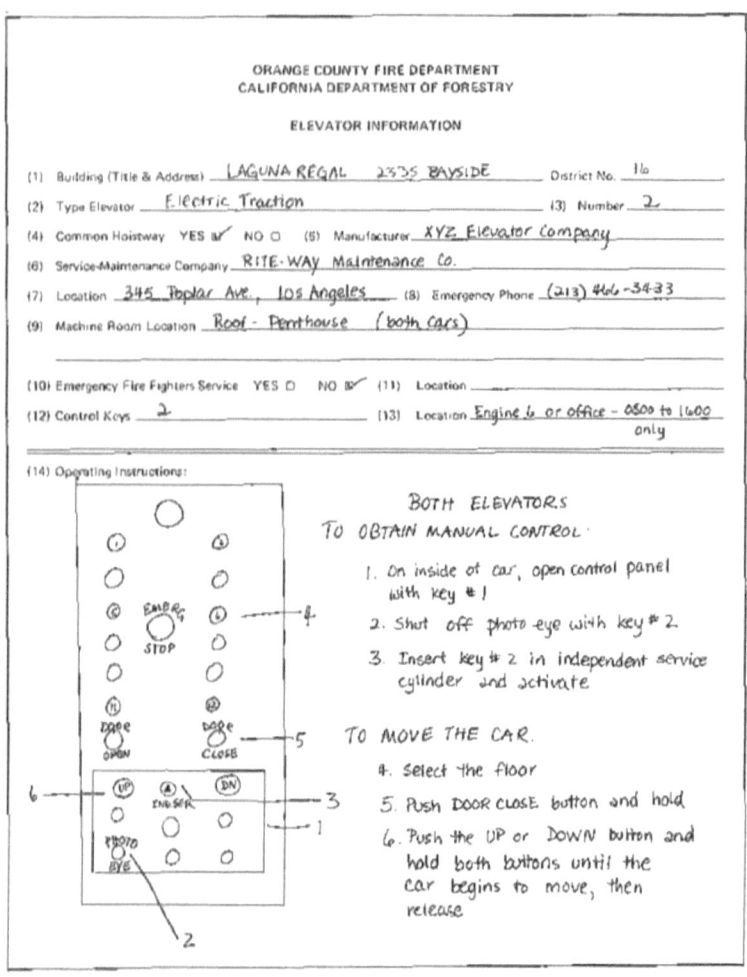

Fig. 2.2. Elevator Information form used by the Orange County (CA) Fire Department.

Operating Instructions (Optional): A drawing of the elevator emergency control panel, accompanied by the steps necessary to manually control the car. Because of space limitations, this information need not be on the same form as the one containing the elevator information, as shown in Figure 2.2, but can be on a separate sheet of paper. Some buildings may have more than one design or type of car; information on each type of car should be included. In such a situation, the car selected for fire operations should be the car that is safest and simplest to operate.

Interlock Tool Pattern (Optional): A sketch or written description of the tool needed to release the interlocks on elevator doors. If necessary, fire service personnel could make a similar tool from this sketch or description, for use during emergencies. This information does not necessarily have to be given on the elevator information form, but can be included on a separate sheet of paper attached to the form.

Special Rescue Problems (Optional): A description of any problems that might be encountered during fire or rescue operations.

Maintaining Prefire Plans

The elevator information form should be kept as up-to-date as possible, should be revised as needed with the elevator emergency preplans, and should be kept with the prefire plans. In their pre-fire plans, some departments include rough sketches of a building and its floor plans, as well as details of the surrounding area in order to illustrate the best approach to the property. Whatever the form used for recording prefire information, it will be more useful if it includes all of the information detailed under "Elevator Information Forms."

Once a prefire plan and diagram have been completed for a particular building, the fire department should maintain contact with the owner or manager to keep informed of any changes that have a direct bearing on the prefire plan. These should include changes in the components of the elevator system and relocation of elevator related items. For example, a firefighter entering the renovated lobby of a preplanned building may find that if the preplans were drawn up before the renovations were made, they do not apply to the new interior arrangement. In such a situation, the time spent trying to find relocated elevator-related items would delay emergency procedures.

CITY OF BOSTON AND COUNTY OF SUFFOLK

DEPARTMENTAL COMMUNICATION

April 14   19 76

	(NAME)	(RATING)	(DEPARTMENT-DIVISION)
TO	John J. McCarthy	Deputy Fire Chief	Planning and Logistics
FROM	William I. Coakley	Lieutenant	" " "

SUBJECT: Temporary Elevator at 60 State Street   FILE REF. No.

The one personnel hoist for this 38-story building is at the corner of
Congress and State Streets. An electrical service to power the hoist is located
in a shed at this same intersection. Of the 3 plywood doors with identical
padlocks, that door which faces Congress Street should be opened to turn on
power to the hoist since it gives direct access to the front of the generator
without entering the shed itself.

The "START" button, colored black, may be difficult to see since it is on
the right side of the generator and not visible from outside the shed. It can
be seen if the operator extends the upper part of his body to the right side while
on the outside of the shed. A red colored "STOP" button is immediately below
this start button.

A repository box with key #2246 is located at ground level of the hoist.
No other key is needed. This key unlocks the padlock to the shed.

PROCEDURE

1. Remove key and unlock shed.

2. Snap up the large master switch on front of generator.

3. Push in the "START" button on right side of generator.

4. Enter hoist and after closing both landing and car doors,
   snap up the toggle on the small red electrical box to en-
   ergize car.

5. For fast lift, push the constant pressure "UP" button.
   For slow lift and when approaching the desired floor,
   the constant pressure button marked "SLOW" should be
   used.

NOTE: It is important NOT approach the upper and lower limits
   of the hoist except in the SLOW speed. An override of
   the floor will cause the safeties to activate and prevent
   further use of the car.

NOTES

1. At this date the floors at which the hoist can stop are: 1-2-7-12-17.

2. A red jewel on elevator panel will light when all electrical contacts
   have been made up . . . (doors closed tightly, toggle switch on).

3. The down direction also has a "FAST" and a "SLOW" speed.

4. Standpipe is up to 11th floor, but only the pipe on Congress St. side
   has connections to every floor.

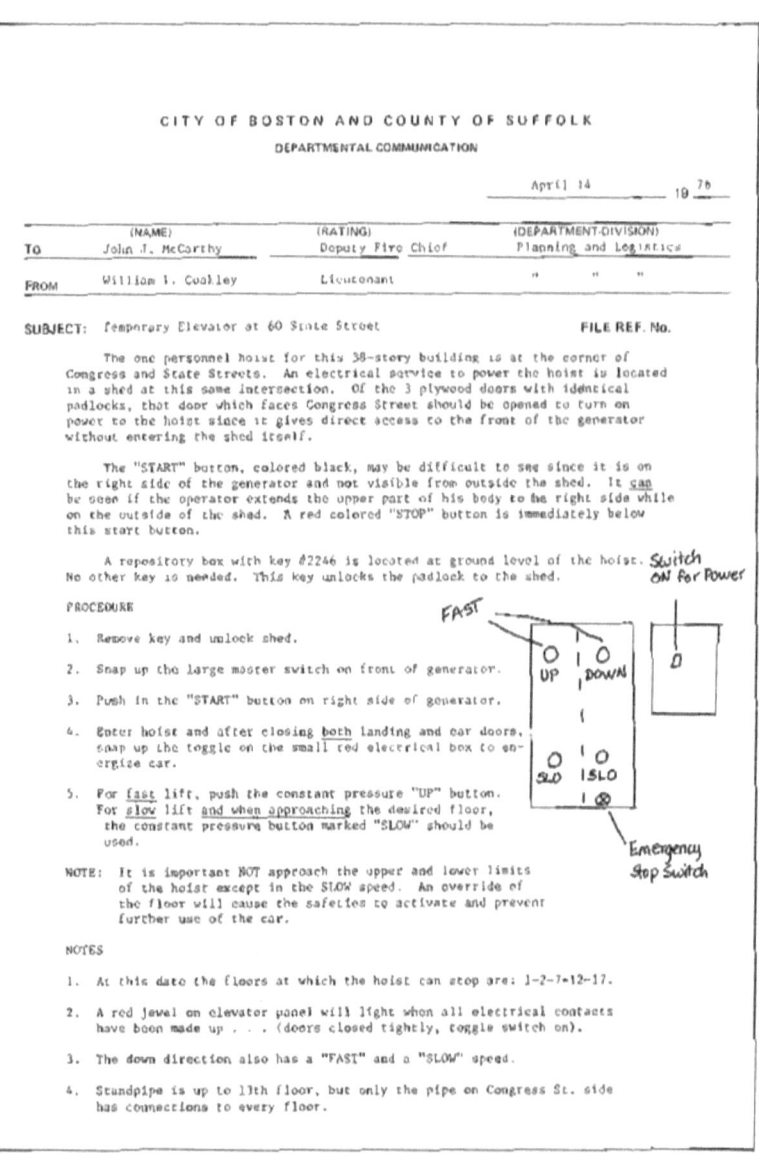

Fig. 2.3. *Information form for a site under construction, with a diagram
and explanation. (Courtesy Boston Fire Department.)*

60

Smoke Detector Installations: Some buildings contain smoke detectors that connect directly to elevator controls. These smoke detectors may affect elevator operation. Some may send all cars to the lobby and some, located in the lobby, may send cars to an alternate floor.

Buildings Under Construction: Just as fire departments must constantly check existing buildings for any renovations that might dictate changes to prefire plans, so must they be aware of buildings under construction in their jurisdiction. Any building under construction with plans for an elevator system will not have a completed elevator installation; sometimes personnel hoists are the only available means for transporting firefighters to the fire scene in buildings under construction. During the day, the hoist will usually have an operator; at night and during weekends, firefighters must be prepared to operate hoists by themselves. Sometimes the power source for the hoist, rather than being close to the hoist, is located in a shack elsewhere on the construction site. Near the power source, or at a convenient, predetermined, on-site location, there should be a set of instructions -preferably with a diagram -for fire service personnel to follow in an emergency. (See Figure 2.3.)

The fire department should be kept informed concerning the progress of the construction, As soon as a temporary hoist is replaced with a permanent elevator, preplans should be revised to include the operational details of the permanent elevator.

Assuring Emergency Services

An important goal of preplanning for elevator emergencies is to ensure that emergency fire service personnel will be properly informed in the event of an emergency. Necessary information includes not only the proper operation of elevators under manual control, but also the availability of keys for the elevator system's Firefighters Service. Any missing keys should be replaced immediately, and deficiencies in the elevator's operation should be corrected.

Older elevators usually do not have the Firefighters Service. Fire departments should contact building owners and inform them of conditions that need to be corrected.

In addition, preplans should be shared with any outlying stations that may arrive first during an emergency. In such instances, outlying towns may have no knowledge or experience in elevator emergencies. Prefire plans, emergency rescue procedures, and training in the operation of elevators should be shared to help ensure that competent personnel will be available when needed.

TRAINING FOR PERSONNEL

Fire departments with elevatored buildings in their jurisdictions should develop training programs that will cover all aspects of emergency elevator operations including the recognition, location, and function of elevator emergency service components. The objective of training firefighters in these procedures is, of course, to enable them to perform as effectively as possible during any elevator emergency. Instructors conducting training programs should be experienced in teaching technical subject matter and should have spent at least 40 hours touring elevatored buildings in their areas. Instructors should also plan to have an experienced elevator company representative present during the training program. Their elevator experience and knowledge will be very helpful to firefighters.

The classroom portion of training in elevator terminology[14] should take anywhere from one to two hours. It should be followed by one to two hours of field work during which the students should examine at least one hydraulic elevator installation and one electric traction installation, including machine rooms and elevator components. Trainees should also view the pit and the top of the elevator car.

Safety is of the utmost importance during field trips. At no time should an instructor attempt to lead a group composed of mare than fifteen students. For viewing purposes, this group of fifteen students should be broken into much smaller groups (possibly three per group) to ensure that everyone will become thoroughly acquainted with the various components. The space in elevator machine rooms, inside pits, and the tops of elevator cars is usually too limited to accommodate large groups safely.

---

[14] See Glossary of Terms at the end of this book.

CHAPTER 3

Emergency Rescue Procedures

INTRODUCTION

Emergency rescues of elevator passengers may be necessary for many reasons. But unless there is a fire in a building, a bomb attack, or an earthquake, passengers in stalled elevators are usually safe if they remain in the elevator car. Simple emergency conditions such as stalled elevators are no reason to endanger the lives of passengers and rescue personnel by undertaking careless and sometimes risky rescues.

As previously stated, it can be extremely dangerous to attempt a rescue of trapped elevator car passengers during an elevator emergency -dangerous both for firefighters and trapped passengers. Except during fires, trapped passengers are usually safer when left within a stalled elevator car. Again passengers are endangered when they leave an elevator car in any way other than walking out normally.

In the first part of this chapter, several procedures for safely handling simple elevator rescues are presented. The second part of the chapter outlines ways to handle crisis situations in elevator installations. These life and death emergency situations demand expert teamwork, skill, and knowledge for safe and effective rescue operations. *As* explained in Chapter Two, "Preplanning for Emergencies," preplanning for elevator emergencies and training personnel to deal with elevator emergencies will greatly reduce incidents such as the following, in which passengers were endangered because firefighters did not know the correct procedures to follow.

Firefighters were summoned to rescue passengers from an elevator stalled just above the fifth floor of a ten-story hotel. Since the car was in a multicar hoistway, they decided to carry out the rescue using an adjacent car. The rescue was carelessly and dangerously undertaken without a man stationed in the elevator machine room, without first disconnecting the power, and without the use of safety lines to secure passengers as they crossed a ladder bridging the hoistway from the stalled car to the rescue car.

In this *case,* the unnecessary and careless removal of passengers through an exit other than the elevator doors could have led to serious consequences had the elevator started up unexpectedly. In addition, passengers and firefighters poorly equipped for safety as well as the absence of an elevator mechanic could have seriously jeopardized the life safety of everyone involved. Such situations can be easily avoided by learning and implementing the rescue procedures presented in this chapter.

PART I - SIMPLE ELEVATOR RESCUE

Elevator mechanics, selected building employees, and emergency rescue personnel should be thoroughly familiar with the correct methods to use when removing passengers from stalled cars. In order to carry out rescue work on a stalled elevator car during an emergency, rescue personnel need to know what procedures to use as well as how the elevator operates and the mechanics of its functioning parts. As cautioned in Chapter One, "Elevator Installations," *only trained personnel should attempt a rescue without the help of an elevator mechanic or qualified manufacturer's representative.*

Whatever the nature of the elevator emergency, a team of trained personnel should be sent to the scene. Metropolitan areas usually have several elevator companies operating in their immediate vicinity. Very often these companies provide 24-hour emergency service.

Many situations are handled within minutes of the time of notification. In rural communities, however, delays of an hour or more are not uncommon. For firefighters, such delays are reason to proceed with a rescue without a representative from the elevator company present. Thus, emergency rescue personnel have greater opportunities to deal directly with simple elevator rescues in these more isolated regions. However, urban rescue teams have no excuse for laxness or complacency about elevator emergencies.

The more elevators there are in an area, the more opportunities exist for incidents to occur which require emergency rescue personnel. For example, during peak periods of use in a large city, massive power failures could disable thousands of elevator cars, entrapping countless passengers. The potential for more routine problems to cause stalling elevators is very real. Elevator cars stop functioning for many reasons. Some of these reasons are loss of power, failure of component parts, and operation of a safety device.

Loss of power may be caused by a general blackout or by a building or elevator power failure. In older installations, malfunctions are more likely to occur where components using electrical contacts are in need of service. Electrical difficulties can include blown fuses, faulty interlocks on hoistway doors, open switches, or breaks in various operating circuits. Elevators contain numerous safety devices to protect the passenger. Unfortunately, these same devices can cause a car to stall between floors.

Again, as a general rule in emergency situations, *it is best to wait for an elevator mechanic* to assist in freeing the passengers as they are safer remaining in the car rather than being hastily removed. The elevator mechanic's knowledge and experience can ensure the greatest safety for the passengers. Often, the elevator mechanic can move the car manually to a landing so passengers can be rescued through the doors. But if a mechanic is not readily available, panic or a medical problem can sometimes compound the emergency. *Only* under extreme emergency conditions should rescue personnel attempt a rescue without the elevator mechanic on the scene. In such cases, emergency rescue personnel must know how to perform quickly and safely the simple elevator rescue operations outlined in this chapter.

ARRIVAL AT THE SCENE

The first step rescue personnel should take when notified of an elevator failure is to make certain the elevator company servicing the elevator has dispatched a mechanic to the scene. Building personnel should be consulted to determine whether or not the elevator company has been notified. If not, the emergency communications center should contact the elevator company while rescue personnel are en route. If the building has no elevator service company or if the elevator company which services the elevator is not available, any reputable elevator company in the area may be telephoned to request that they send an elevator mechanic to the building. During preplanning, local elevator companies should be contacted to find out if they will assist in an emergency that does not involve equipment they have either installed or maintained. As a last resort, the first-due officer can notify the elevator company upon arrival at the building. This assures rescue personnel that expert advice will be available, if not immediately at the scene, then at least by phone. In some areas, this notification is required by the elevator and/or building and safety codes.

Dispatching one squad or a truck company and an engine usually assures adequate personnel and equipment to handle the problem safely in the least amount of time. It is imperative that rescue personnel reach the scene as quickly as possible. Panicky passengers or unreasonable bystanders can complicate simple emergencies sometimes with tragic results. For example:

> A passenger who would not wait to be rescued tried to exit through the partially opened door of an elevator stalled with approximately a foot of clearance between the elevator floor and the hoistway entrance ceiling. As he prepared to step down to the floor landing, the car suddenly moved, pinning him against the ceiling, thereby killing him. Two people who tried to help him had their arms caught when they attempted to prevent the elevator from crushing the victim. The subsequent rescue operation to free the would-be rescuers took hours. Fortunately for all involved, no one fell into the open area under the car to the pit floor, a very real danger in this incident and others like it.

When rescue personnel arrive at the scene, one person only should supervise the rescue attempt. Otherwise, problems may occur as building maintenance personnel or bystanders attempt to give orders. Rescue personnel must follow only the instructions of one individual as orders from unauthorized persons could unnecessarily and even tragically complicate the rescue operation.

COMMUNICATIONS WITH PASSENGERS

In all emergencies the rescue personnel's first consideration and responsibility is to reassure and convince the passengers that they will be rescued. Using the elevator's communications system, one rescuer should identify himself and reassure passengers that immediate steps are being taken for their release, that they are safe, and that they should not smoke. The rescuer should find out whether any passengers are ill or injured, how many people there are in the car, whether the lights are on, and where the car is located. Next, the rescuer should tell passengers to stand by the Emergency Stop Switch and that they will be safe as long as they do not panic, touch any controls, or try to open the doors without specific instructions from rescue personnel. If the elevator car has no direct communication system, a member of the rescue team should locate the elevator in the hoistway and reassure the passengers from the nearest hoistway door. Since unruly or uncontrolled passengers who might also be on the verge of hysteria may seriously hamper any rescue attempt, a strong forceful approach is often required.

Contacting the passengers to gain control and ask questions may be one of the more difficult tasks. Two systems are usually available for communicating with passengers: an intercom or a telephone. A two-way intercom, if one is available, is the quickest way to communicate with trapped passengers. Usually, it connects the car to the main lobby and is found on the main control panel. Most elevator codes require some means of communication in modern elevator cars. Direct line telephones are usually the best way to deal with the passengers. However, if there is no direct line to the elevator car, it is best not to relay instructions through telephone operators. Omitted information could easily complicate the problem.

When a system for direct communication to the elevator car does not exist, a member of the rescue team may yell or call up the hoistway to the passengers. Outside distractions from heating, air-conditioning, or other cars may need to be controlled, however, before any conversation can take place. Rescue personnel should continue to maintain contact with passengers during rescue operations until they are safely out of the elevator. This will reassure them of their safety and the progress of the rescue.

LOCATING THE CAR

Often, the stalled car will have been located before firefighters arrive on the scene. In situations where the car has not been located, there are several methods that may be used to determine its position in the hoistway.

1. At the lobby floor, observe the position indicator light (if one is provided). It may show the floor nearest to the stalled elevator. However, if the car is between floors, the lights may not be lit.

2. Web-Based technology is becoming more and more common. if a building has such capability and the power is on the building maintenance personnel can provide you with all kinds of information including the exact car location.

3. Fire Control rooms found in some high-rise building generally have elevator monitoring equipment that will provide car location information.

4. Ask the passengers to give the floor indicated on the interior car control panel.[15] Since the panel displays upcoming floors as it travels in the hoistway, it too will show the floor nearest to the stalled elevator. This method is only useful when communications have been established and there has not been a complete power failure.

5. 'When adjacent cars are operating in a multiple hoistway, ride up in the nearest elevator, stopping at each floor. Look across and upward through the narrow space between the car and hoistway. You will need a light for searching the hoistway in this way.

6. If all else fails and you are sure you have the right elevator and hoistway, open the main line disconnect switch

---

[15] This may not always be correct especially with express elevators or elevators that operate at very high speeds. Some times the difference in high speed elevators can be two to five floors.

68

and then open the hoistway door at the lowest floor in order to look up the hoistway. 'When an elevator has been stalled near the top floor of a taller building, the location of the car can be learned from observing the position of its counterweight if the system is an electric traction one. If the car is near the top of the hoistway, you will be able to see the counterweights.

ELEVATOR RESCUE WITH POWER ON

There are many reasons why a car could stop when there has not been an interruption in the power supply. Regardless of the reason for the stalled car, it should be standard policy to instruct the passengers to make sure that the interior car doors are closed.

Once rescue personnel have talked with the passengers and located the car, one member of the team should go to the machine room to stand by the main line disconnect switch. Generally, machine room breaker switches are in direct sight of the selector and hoisting equipment. These devices are usually the fuse type and should be checked by the elevator maintenance man or building superintendent. The preferred safe practice when removing trapped passengers is to move the elevator to the lowest level. This procedure must only be attempted by an experienced elevator mechanic. Rescue personnel should never try to move an elevator.

Weak or dirty electrical contacts can cause elevator cars to stall when there has not been an interruption of the power supply. If this is the case, go to the closest floor landing and instruct a passenger to push either the DOOR OPEN or FLOOR button on the car panel. At the same time, you will push and hold the hall button. This action may open the doors if the car is within the landing zone. This should only be done with the rescue team present at that landing.

One of the most common reasons for cars to stall between floors is open electrical contacts at the hoistway interlock. Interlocks can become weak and dirty. Should this happen, movement of the car can cause the points to lose contact and immediately stop the car. To correct this problem, the following procedure is recommended. With the power on:

1. Have the passengers shake the interior car door.

2. Have rescue personnel shake the exterior hoistway door nearest the car's location.

3. Have rescue personnel shake the exterior hoistway door where the passengers entered the car.

4. Have rescue personnel shake each hoistway door at all the landings. If building height makes this procedure impossible, shake the hoistway doors between the point where passengers entered the car and its present location.

If a car is in the landing zone, the door may open. In other cases, the car may proceed to the next floor where a call has been entered and the interior car door will open, freeing the passengers.

Should an elevator car be in the landing zone (within floor range), the door locks may have already been released. In such cases, merely pulling the doors by hand can sometimes open them. In some cases, such as when the car is stalled in the landing zone, it may be possible to open the door by pushing the hall call button or by breaking the light beam from the photoelectric eye if the car is equipped with one. Again, with the power on, insert a thin but stiff piece of paper or cardboard between the door and the door jamb to break the beam, as shown in Figure 3.1. In some cases, this method may cause the door to open. This should only be done with the rescue party present at that landing.

ELEVATOR RESCUE WITH POWER OFF

If rescue personnel cannot free passengers with the power on, the incident commander needs to reassess the situation to determine how to continue rescue attempts. Remember that in simple elevator rescues that do not involve fire, etc., the passengers are safe as long as they remain in the stalled car where there is little possibility of injury. Their safety is always the primary concern during the rescue.

Securing the Car

Since there is no guarantee that the stalled car will not start again, all possible safety circuits need to be opened to ensure against accidental movement. Opening the main line disconnect switch to the elevator will cut off power to it and to the brake coil. Main line disconnects are found in the machine room which controls the stalled car. When the power is cut off at the main line disconnect, the lighting and ventilation systems should remain operational as almost always they are connected to independent electrical circuits.

One member of the rescue team must remain stationed at the breaker to ensure that no one reactivates the power until so directed by the person in charge. Two-way communication between the rescuer at the main line disconnect switch and the incident commander is required. Serious injury could Occur if power is left on or restored prematurely while passengers or would-be rescuers are attempting to enter or exit the stalled car. As an additional precaution against accidental movement of the stalled car, the passengers should be told to make certain the Stop-Run Switch or Emergency Stop Switch (usually colored red and labeled) has been *set*. Often there are operating instructions either on the switch itself or beside it. Authorities recommend that both switches be set before attempting rescue operations, provided an alarm is not activated which could complicate communications.

Before removing passengers from an electric traction elevator, it is necessary to check the hoistway to make sure the elevator cables are not slack. Slackness often means that the car may move at any time.

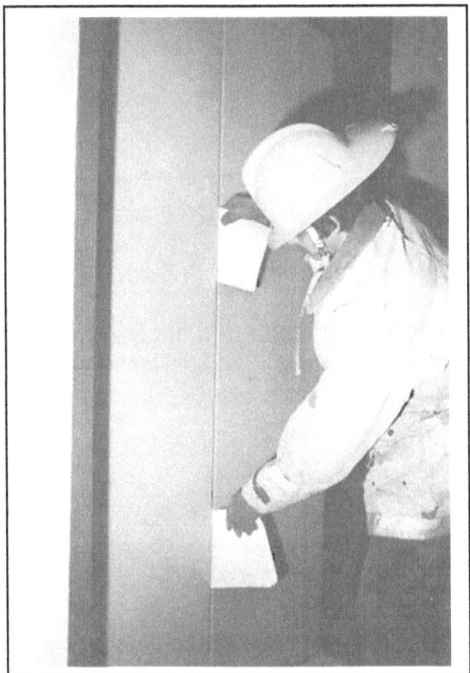

Fig. 3.1. *Inserting a thin but stiff piece of cardboard to break photo beam and open hoistway doors.*

In such a situation, the car should be supported, hung, Or blocked into position before removal of passengers *is* attempted. The same is true of any elevator car (hydraulic or electric) where the floor or ceiling of the car is tipped at an angle.

Opening Hoistway Doors

Once the car has been immobilized, rescue personnel can proceed to free the passengers. The preferred method *is* to remove trapped passengers through the hoistway doors. This means the hoistway

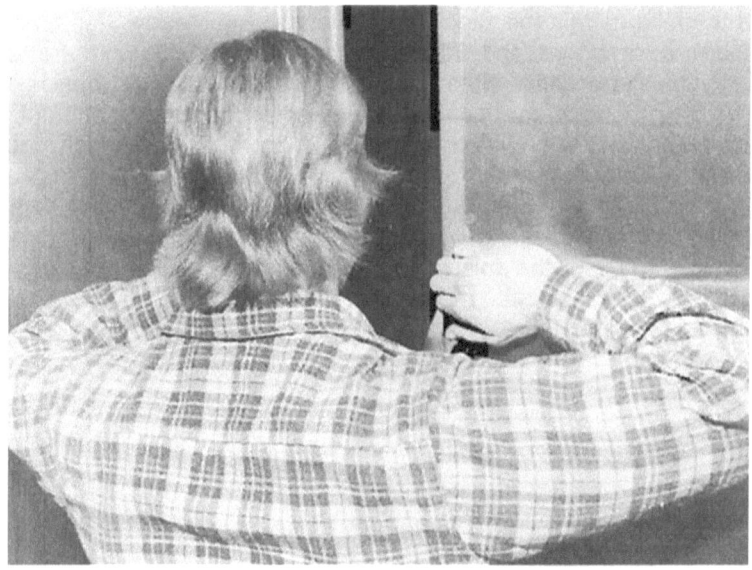

Fig. 3.2. Passenger *pulls on car doors in order to open them.*

door must be unlocked and opened. Since there are several ways to open hoistway doors, rescue personnel should be familiar with all of them and should follow sequential steps from least to most difficult until the doors open.

Passenger Assisted Entry: In many situations, passengers can release themselves from a stalled car since doors can often be opened by hand. With the power off, instruct the passengers to push or pull the car door open, as in Figure 3.2. This will require some effort to overcome the effects of friction on the door. Since the movement of the interior door will trip the interlock on the hoistway door, rescue personnel working from the landing should be applying hand pressure to the hoistway doors in the direction that they travel. (See Figure 3.3.) This procedure may open both doors of the car if it is within the landing zone (Generally within 18 inches of the landing). If the car is outside of the landing zone and not equipped with a door restrictor, (See Chapter 1 , page 35 door restrictors) only the interior door will open. In this case, it will be necessary to instruct a passenger to move the pickup roller or interlock link on the hoistway door to release its interlock and open the door. It should be explained to passengers what the interlock looks like, where it is located, and how it works. Tripping the interlock requires very little pressure, but again, exterior hand pressure from the landing side will help in opening the door. The hoistway door should be opened *slowly,* for suddenly opening the hoistway door may cause elevator passengers to lose their balance and possibly fall out of the elevator. If the interlock is out of reach, passengers will need a tool to help them reach it. On an elevator with center opening doors, a slim tool such as a yardstick or a pry bar can be inserted between the doors.

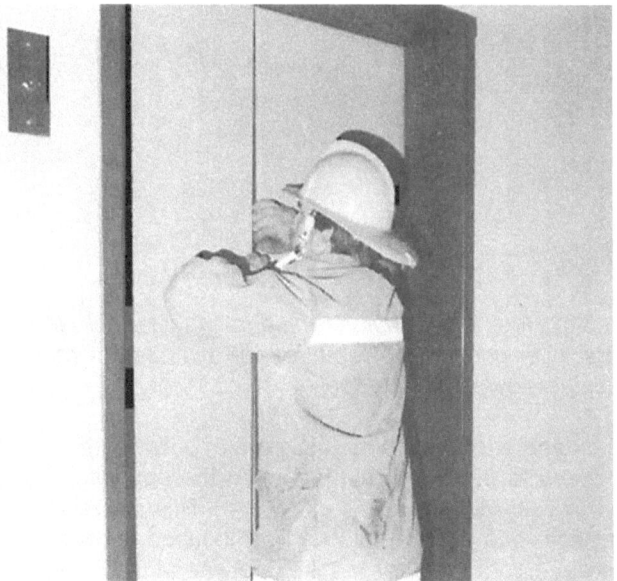

Fig. 3.3. *Fire fighter pulls on hoistway doors to open them.*

In some cases, the passengers may be able to open the interior car door a limited distance by pushing or pulling, but only with the power

73

off and the rescue party at the landing. If this should happen, it usually indicates that the power door operator, pickup, or clutch assemblies are not functioning correctly.

Use of Tools to Gain Entry: If none of the procedures described thus far has been successful, it may be possible to trip the interlock from an adjacent elevator car in installations having multiple hoistways. By running an adjacent car up to the level of the stalled car, a rescuer

Fig. 3.4. Here a pike pole is used to trip interlock and open hoistway doors. (Car doors were removed for this photo; ordinarily they would block the view of the interlock.)

can extend a tool (such as a pike pole) between the hoistway and the car doors to trip the pickup roller on the opposite unit, thus opening the hoistway doors. (See Figure 3.4.) This is usually a three-person operation: one person hold~ a light focused between the car and the hoistway doors, another reaches between the doors with the tool and trips the interlock, and a third person opens the hoistway doors from the landing. This should only be attempted with the power off for the stalled car, rescue car, and all cars adjacent to the rescue car.

Most states permit emergency unlocking devices which are operated by a special key-type tool. These devices are inserted through the hoistway door via a hole, tripping the interlock mechanism. (See Figures 3.5 and 3.6.) There are several types of unlocking devices, each of which operates by applying pressure to the pickup roller, thus releasing the hoistway door locking device. Rescue personnel operating in areas where codes have not excluded the use of these tools should either carry the necessary tools or make arrangements

Fig. 3.5. (Above) Pointer shows the keyway in which a key may be inserted to trip the interlock and open hoistway doors. Fig. 3.6. (Next Page) Here a rescuer uses one kind of unlocking device to release center-opening doors. (Photo courtesy of Otis Elevator)

Exterior view side exit
(Photo courtesy of
Elevatorbob Web Site)

*Fig. 3.7. In this picture the two elevators are side by side,
and the emergency side exits of both are open.*

with building management to have such tools available for emergency use. Excessive pressure is not necessary when using interlock tripping tools. It is a matter of knowing how and where to insert the tool.

Rescue personnel should inspect a car for lock locations and should have prior training in the *use* of these tools. As a general rule, an elevator car's releasing rollers move away from the leading edge of a door.

For hoistway doors which close toward the adjacent car, the roller is pushed away to release the interlock. For hoistway doors which close away from the adjacent car, the roller is pulled toward it, tripping the lock.

Biparting and hall swing doors usually do not require a special tripping device because such doors have glass vision panels located within easy reach of the interlock mechanism. By simply breaking the glass vision panel, the hoistway door may be opened with very little damage or effort. This should only be done with the power off.

Entry Through Emergency Exits: In situations in which rescue personnel cannot trip the locking mechanism from an adjacent hoistway or the passengers cannot carry out procedures to release the interlock from inside the car, it will be necessary for a rescue team member to enter the car. Entrance to the car is usually through the top escape hatch or by the side emergency exit. These will be equipped with electrical contacts to shut down power to the elevator.

As noted in Chapter 1, elevators operating in multiple hoistways have emergency side exits. Most side exits are at least 16 inches wide and 5 feet high. (See Figure 3.7.) Entrance to a stalled car in a multiple hoistway may be gained through this side exit by way of the side exit of an adjacent elevator. When an operable adjacent car is brought to the same level as a stalled car, the power to both cars should be shut off to prevent any accidental movement. Once the side panel is opened, electrical contacts will open a safety circuit, providing additional protection. Side exits are either hinged or removable panels; however, neither type opens easily. The hinged panel is opened from the inside of the car by means of a removal key, which should be readily available for use by emergency rescue personnel. The side exit of a stalled car is opened from the outside by means of a non-removable handle or handles. The trapped passengers should be warned to stay clear of the door, which will open inward. Once the side panels on both cars have been opened, an emergency evacuation bridge can be placed across the hoistway between the cars. The rescuer who enters the stalled car should check to see that the car's Emergency Stop Switch is in the STOP position. He then should assist passengers from the stalled car to the rescue car. Safety belts and lifelines must be used by rescuers and passengers when crossing from one car to the next.

Top escape hatches usually open outward and may be removable or hinged. They are secured in the closed position and usually can be opened from the top of the car only. On older installations, top panels can usually be removed from either the inside or the outside of the car. *Access* to the top of the car is gained by opening the hoistway door from the floor landing above the stalled car. From this landing, a member of the rescue team should be lowered onto the top of the stalled car. The Stop Switch at the mechanic's top of the car station should then be activated. "When rescue team personnel arc working in a hoistway, they should use lifelines to prevent accidental falls. After opening the top escape hatch, the rescuer may need to remove the ceiling panels often found in modern elevators. These panels may sometimes hinge and swing down into the car. Thus care should be taken to remove the panel, which can usually be slid aside, so passengers are not injured. Once inside the car, the rescuer should be certain that the Stop Switch is set.

Forcing Entry: When all else fails, the rescue personnel may pry or cut open the hoistway doors. If this method is chosen, it is recommended that whenever possible rescue personnel wait for a qualified elevator mechanic to assist them. Hoistway doors are extremely costly to replace or repair; however, they can be forced open with minimum damage when care is taken to apply force at the proper point. If rescue personnel must pry the door open, force should be applied as close to the interlock as possible. The interlocks on automatic elevators are usually found near the top of the door on the jamb side for single-speed or two-speed sliding doors. For center opening doors, the interlock is near the top of the door where the two doors meet. A good tool to use is a low-pressure spreader, since high-pressure spreaders will do too much damage to the doors. Sufficient force applied close to the interlock, will either cause it to break or the door to fold. Excessive force is not necessary and can result in costly damage. Knowing where to apply force can help prevent such damage.

In one situation which occurred in greater Los Angeles, firefighters used pry bars and a spread ram to apply force one half to three quarters of the way down from the top of the door. The door had already been damaged there by building occupants trying to rescue a trapped victim, an elderly man with a heart condition. After the force was applied to the door, the man was released in 20 to 25 minutes. The damages to the elevator amounted to a few thousand dollars. Damages would have been less costly if force had been applied at the top of the door. They would have been minimal had the firefighters been familiar with door construction and where to apply force.

Fig. 3.8. *When a car* is *stalled more than* 3 *feet below a landing,* it is *recommended that passengers be removed through the top escape* hatch.

Removing Passengers

Through the Doors: After the hoistway and interior car doors have been opened, rescue personnel can remove the passengers. Depending on its location when the failure occurred, the stalled car will be flush with or close to the landing, above it, or below it. When the car is stalled near the landing, removal is safe and easy) and in some cases passengers can walk out. Before allowing passengers to exit, a rescuer should enter the car to see that the Emergency Stop Switch is set, provided it does not cause a communication problem due to a bell alarm. Another rescuer should stand on the landing and assist exiting passengers. Assistance is especially needed when the car is not aligned with the landing. This distance could be as great as 18 inches.

When an elevator car has stalled more than 18 inches below the landing, passengers will need to use a ladder to climb up and out of the elevator. (See Figure 3.8.) Before they exit, one rescuer should enter the car and check to see that the Emergency Stop Switch is in the STOP position. That rescuer then steadies the ladder in the car while another squad member braces it at the landing. Both rescuers should then assist the passengers as they climb up the ladder and out of the elevator car onto the landing.

Fig. 3.9. When a car is stalled above a floor landing the open hoistway should always be blocked. Otherwise, a passenger or rescue worker could fall into the pit, and unnecessarily complicate an already difficult procedure.

A reverse situation exists when the elevator car stalls more than 18 inches above the landing. As with the procedure for a car stalled below a landing, one rescue team member enters the car and makes certain the Emergency Stop Switch is in the STOP position. He then helps each passenger onto the ladder while another rescuer at the

80

landing steadies the ladder and assists passengers down to the landing. Before ,removing anyone from a car that is stalled above the floor landing, make certain that passengers cannot fall into the open hoistway. This has happened with tragic results.

In one situation, a 34 year old woman was trapped in an elevator with two other persons just above the sixth floor landing of a residential condominium building in Alexandria's Chinatown. One man, an FBI agent, got out of the elevator and when the panicky victim tried to follow him slipped falling to her death. This was a 21st-century elevator according to the article. [16]

In another situation, a 75-year-old man fell 12 stories down an elevator shaft to his death after trying to exit an elevator that was stuck between floors. He was the last person to evacuate the stuck elevator when he slipped, lost his balance and fell down the shaft according to the president of the condominiums board where the accident occurred. [17]

To prevent a similar situation, the hoistway opening should be blocked by a ladder or a plank, as shown in Figure 3.9. Should the length of the door opening be limited in this situation, rescue is complicated by the increased danger of people falling into the enlarged opening to the hoistway. Rescuers guide and assist passengers down the ladder or plank until they safely reach the landing.

Through Top and Side Emergency Exits: When removing passengers through the doors is impossible, it is necessary to use one of the car's emergency exits. As described earlier, these are the emergency side panels and the top escape hatch. Which exit to use will depend on the condition of the passengers and the location of the car in the hoistway. Injured, ill, or elderly passengers may be unable to climb through the top escape hatch; however, if the car is too far above the floor landing, or located in a single hoistway, or if an adjacent car is unavailable, exit through the top escape hatch may be the only choice.

The earlier section of this chapter titled "Opening Hoistway Doors" described how to gain entry to a stalled elevator through the

---

[16] Washington Post 2005
[17] Associated Press 12.22.06 Houston, Texas

emergency side panel. Follow those same steps until a rescuer has entered the stalled elevator car. The rescuer should take the usual precaution to check that the Emergency Stop Switch is set before assisting passengers through the emergency side exit and onto the emergency evacuation bridge leading to the other car. For additional safety, passengers should be secured with a safety belt and lifeline. Another rescue team member waiting in the adjacent car should assist the entering passengers. After the passengers have been transferred, the emergency evacuation bridge and any other equipment should be removed, the emergency side exits replaced and locked, and the rescue elevator returned to the main floor. Make certain to leave the power off for the stalled car.

Fig, 3.10. (Above) The top escape hatch of a stalled elevator has been removed and a nonskid ladder has been securely positioned inside car, Fig 3.11. (Right) Lifelines should be attached to both rescuer and passenger before removal is attempted.

To remove passengers through the top escape hatch, a member of the rescue team enters the elevator by following the procedures for gaining entry outlined in the earlier section of this chapter titled "Opening Hoistway Doors." Once the escape hatch is opened, a ladder with nonskid safety feet should be let down into the car and securely positioned. (See Figure 3.10.) The ladder should extend three feet above the top of the elevator. The rescuer enters the car, checks that the Emergency Stop Switch is in the STOP position, and remains in the car to secure passengers to a lifeline and assist them up the ladder onto the top of the elevator car. (See Figure 3.11.)

Fig. 3.12. (Left) A rescuer *should assist passengers* at *all times during the rescue. Fig.* 3.13. *(Above) Passengers should be helped one at a time out* of *the car and hoistway onto the landing.*

Another member of the rescue team stands on the top of the car, secured by a lifeline, to help the passengers one at a time through the opening. (See Figure 3.12.) If still another ladder is needed to reach the landing, a rescue team member may be stationed on top of the car to help passengers. This ladder should extend three feet above the hoistway landing. One more rescue team member should be stationed on the floor above the car to assist passengers off the ladder and onto the landing. (See Figure 3.13.)

ELEVATOR RESCUE IN A BLIND HOISTWAY

There are situations where elevators become stalled in blind hoistways. Since rescue is often difficult and dangerous, the rescue team may prefer to wait until an elevator mechanic is available to assist them. Some shafts may be blind for 30 floors or more, making location of a stalled car extremely difficult. In a multiple hoistway, rescue personnel can sometimes use an adjacent car to search the hoistway and locate a stalled car by stopping the car and looking out from the side emergency panel or the top escape hatch. Utmost caution is needed when looking out from a moving elevator car. When the rescue car is stopped close to the stalled car, a rescue team member should open its side emergency panel to determine the position of the rescue car in relation to the stalled car. He then pushes the floor call button for the floor nearest to the stalled car. Since the rescue car will not move unless the panel is closed, it can be moved by opening and closing the panel so that the electrical contacts on the panel meet. The rescue car can be moved each time the panel is closed until it is aligned with the stalled car. An elevator car should never be operated with the side or top emergency exit in the open position. By moving the rescue car alongside the stalled car, passengers' can be removed through the side emergency panel, as described in the previous section.

Should an elevator car be stalled in a blind hoistway because of a power failure, adjacent cars also affected by the power failure, cannot be used to locate and rescue passengers. In such cases, passengers can sometimes be located by sound. For example, using the emergency powered intercom, a rescuer should ask a passenger to tap the car wall so that other rescuers listening at each floor can determine the car's location.

If the blind hoistway is in a single shaft, there will be access doors located at approximately every third floor. These access doors can be used to locate and remove passengers. Once rescue personnel have located the car, passengers can be removed from the elevator through the top escape hatch. In extremely critical cases, hoistway walls may need to be breached in order to rescue passengers from a blind hoistway in a single shaft.

NOTE: When emergency power is available, building services personnel may be able to move stalled elevators, usually one at a time.

PART II - COMPLEX ELEVATOR RESCUE

Unfortunately, not all elevator emergencies are simple some 200 people die each year in the US because of elevator mishaps.. Complex elevator emergencies, other than those involving fire and natural disasters, occur in which people may be critically injured or in serious danger of losing life or limb, emergency rescue personnel may be endangered, and the time element is critical. In most cases, complex elevator emergencies happen in older installations which do not have the modern safety devices found on newer cars. In these situations, teamwork, skill, and knowledge are necessary for a safe and effective rescue operation.

Most complex elevator emergencies involving rescues are usually of four types. By far the most common are those where the victim falls into the open hoistway. In other situations, the victim is pinned between the car floor and the hoistway door opening. This situation occurs when a car suddenly ascends while a person is stepping through the car doorway. Another type of emergency occurs when the victim is pinned between the top of the car doorway and the hoistway floor landing. This usually happens when the car suddenly descends while a person is entering or exiting through the car's doorway. Lastly, some accidents can pin a victim between the car floor and the hoistway landing.

BASIC PRINCIPLES FOR HANDLING

COMPLEX ELEVATOR EMERGENCIES

Although there are many types of elevator installations operating under a variety of mechanical conditions, there are a few basic principles universal in all serious emergency situations involving life and death.

Because installations vary in age and type, it is essential to have the advice of a trained elevator mechanic. Sometimes, however, it is impractical to wait for a mechanic to arrive on the scene. In those situations, either the mechanic or the elevator company that installed or services the car should be contacted by phone for advice. Their recommendations should be followed if at all possible.

The first consideration in complex elevator rescues must be to prevent further injury to the victim or injury to rescue personnel, bystanders, or even passengers in adjacent elevators. Precautions should be taken to prevent the victim from falling into the open hoistway, the car from moving accidentally, and the possibility of further injury from rescue equipment. Of equal concern is the victim's medical needs. Where necessary, first aid should be provided immediately upon arrival. The victim should be supported to ease pain, reduce shock, and prevent any major loss of blood. The area should be well-lit, including the areas above and below the victim.

Until a definite plan for removal is established, do not permit unnecessary car movement. Unintentional movement could result in more serious injury, jeopardizing the victim's life. Sometimes material that has been pressed into or pierced the victim's flesh compounds the problem of removing the victim. In cases where the material has been pressed into the victim, it needs to be removed, preferably unabrasively in one clean movement away from the point of body contact, without twisting or cutting. Material which has pierced the victim's flesh needs to be cut away from the victim. Once the material has been removed, severe bleeding may occur, but medical or paramedical personnel at the scene may be able to prevent this from happening. In any case, the rescue team should be prepared to administer first aid.

Rescue personnel need. to know the potential hazards they face when dealing with complex elevator rescues involving injured people and, although speed is essential, all rescue operations should be performed with the utmost safety precautions. For example, rescue personnel should wear clothing that will not impede rescue operations. Loose clothing and long-sleeved shirts are dangerous around elevator equipment. Moving object~, hoistway projections, and the moving equipment of an adjacent car all present hazards to rescue personnel, who should be well-aware that such potential dangers can ultimately affect the success of the rescue operation.

EMERGENCY RESCUE OF PINNED VICTIMS

Before attempting to rescue a pinned victim, the rescue team needs to determine an appropriate strategy for dealing with the emergency.

Generally, there will be three methods to choose from, two of which involve moving the elevator car: [18]

1. Moving the car up or down, away from the victim.

2. Forcing the car directly backwards away from the doorway and into the hoistway, away from the victim.

3. Anchoring the car and cutting a portion of either the car, hoistway wall enclosure, or building floor around the point where the victim is trapped.

There is usually no choice from among the three methods. which one is used depends on the way the victim is pinned, the design of the elevator unit, and the construction of the building.

All necessary safety precautions need to be taken before any rescue operation begins. The car needs to be immobilized as described in Part I of this chapter, "Elevator Rescue with Power Off." The power should be disconnected at the main line disconnect and a rescue team member stationed at the switch to be certain it is not turned on again. As an additional precaution, power to adjacent cars should be turned off to prevent additional accidents. Usually, disconnecting the power at the main line disconnect switch is a sufficient safeguard against any car movement. When a person is pinned between the car and the hoistway, the possibility of the car's sudden movement is always present, even with the power off. This is especially true on electric drum-type elevators supported by cables where the cables on the winding drum may have slackened from the victim's presence. Once the victim is released, the car will move to take up the slack. As an additional precaution against any sudden movement of the car, the use of sturdy props or wedges between the car and the hoistway, or lash ropes tied to structural building columns and the car, is recommended.

Securing the car in the preceding manner allows rescue personnel to apply the first method; i.e., carefully move the car up or down in the shaft, thus safely freeing a victim who is only slightly wedged. Wood props and wedges as well as rope lashings allow the car to be moved

---

[18] NOTE' It is imperative that moving the car up or down in the hoistway be done by, or under the direct supervision of an elevator mechanic at the scene. Forcing the car backward, or anchoring the car and cutting where necessary, should be done only under the supervision of an elevator mechanic unless waiting for his arrival threatens life or limb.

in the right direction and give the rescuer control over the distance the car moves. Before wedging is attempted, the hoistway above the car should be examined for slack cable, which could suddenly be taken up when the car is moved free. when the machine room is at ground or basement level, the slack rope may be on the return side so as not to be visible in the hoistway above the car. With the supports in place, the car should be carefully moved until the victim is released. The car may also be moved by its normal hoisting equipment if an elevator mechanic is available to closely supervise the operation. Another method for moving the car away from the victim is to use jacks or spreading equipment between the car and the floor landing. The best equipment for this purpose is a high-pressure spreader or hydraulic jack. When high-pressure spreaders are used, extreme caution is necessary to prevent the building up of excessive pressure. Too much pressure could cause the car to shoot rapidly upward or downward out of control. Safeties can also become a hazard. There are more than a dozen kinds of safeties, many of which can be released by lifting the car or its counterweight, or both. Other kinds have more explicit release procedures. There is not always sufficient time for rescue personnel to determine the kind of safety in use, much less the specific procedure needed to release it. And needless to say, the use of spreading equipment prying upward could lead to serious results. Again, it is imperative that the advice of an elevator mechanic be obtained before attempting this or any other method of moving an elevator car.

There are situations when the second method -forcing the car back into the hoistway is the best way to free a pinned victim. In most situations, the car may be forced back into the hoistway by removing the upper guide shoes. First, the top guide shoes (which are usually bolted to the crosshead) should be removed. Then, enough pressure should be applied by using spreading equipment or jacks in order to gain sufficient clearance to free the victim. Another way to force the car back into the hoistway involves cutting the guide rails above and below the lower guide shoes. Still another way to force the car back into the hoistway would be to apply sufficient force to tear the guide rail brackets from the hoistway wall, thus freeing the victim. Low pressure spreaders and hydraulic jacks can be used for this. These methods are both dangerous and difficult and should only be attempted with the aid and advice of an experienced elevator mechanic, and then only in an extreme emergency. In some cases, it may be possible to cut the guide rails on one side only, enabling the car to pivot on the rails on the other side (similar to the action of a swinging door), thus freeing the victim. However, this procedure will invalidate the car safeties. Any time it is decided to force a car back into the hoistway, there must be sufficient clearance to make the operation feasible. The elevator mechanic on the scene must make the final decision as to whether the procedure can be safely accomplished. One additional note of caution -before attempting a rescue maneuver of this type, it is recommended that the car be provided with additional support before any backward movement is attempted.

The third method for dealing with the emergency rescue of pinned victims involves cutting the material from around the point where the victim is trapped. This will mean cutting either the car, the hoistway wall, or the building floor. Usually, the best method is to remove a portion of the hoistway wall enclosure, and then cut or pry the door frame to free the victim. If this method is not possible, try cutting and prying portions of the car away from the victim. It is advisable to use limited sparking equipment such as an air hammer or metal hand saw when cutting. Charged hose lines and/or extinguishers should be available. When the floor around the victim is wood, it may be easiest to cut away some portions of the flooring. Always use caution, and support weakened areas from below if it is necessary.

RESCUE PROCEDURES FOR SOME

TYPICAL ELEVATOR ACCIDENTS

Training personnel to handle complex elevator emergencies can be complicated by the wide variety of elevator installations and building construction features. Therefore, rather than using standardized training procedures, it is sometimes helpful if hypothetical emergency situations are developed, applying some of the points in this chapter to these situations, and using local installations to work out the situations. This would tailor training sessions to the needs of your community. Student discussions should be elicited in order to encourage group problem solving of some of the situations. Group problem solving offers the opportunity to present alternative methods of dealing with a particular problem. In any training situation, it is important to reinforce the need for handling elevator emergencies with the assistance of a qualified elevator mechanic, and to emphasize the need for safety precautions.

The following hypothetical situations illustrate the four most common types of complex elevator emergencies and their suggested rescue procedures. In each case, the suggested rescue procedure is outlined in a series of steps. You may find it useful to adapt these procedures to meet your particular needs, remembering that in any serious elevator accident, it may be necessary to administer first aid.

Situation I: Victim Falling into the Hoistway

As mentioned earlier, the majority of complex elevator emergencies involve persons who have fallen into open hoistways. In most cases, this is due to human error rather than equipment malfunction. Often, the victim is a building employee who has carelessly opened the wrong hoistway door. The following procedure is suggested for this type of situation:

1. Use the main line disconnect switch in the machine room to cut off all electric power, and the Emergency Stop Switch in the elevator car to ensure that the car will not move.

2. Locate the victim. (The victim may have landed on the car top or in the pit.)

3. Open the hoistway door nearest the victim by one of the methods outlined in Part I of this chapter, "Simple Elevator Rescue." Remember to block the hoistway opening to prevent anyone from falling into the pit. When the victim is in the pit, rescue personnel may be able to reach him from a door that leads to the pit area. Otherwise, a ladder will need to be let down from the lowest landing, again blocking the hoistway landing to ensure against accidental falls. Before entering the pit to rescue the victim, throw the Pit Emergency Switch to the OFF position.

4. If the victim has fallen into the hoistway and landed on the top of the elevator car and is wedged between the car and the hoistway wall, use the procedure outlined in the following Situation II. When the victim has landed on the top of the elevator car and is not wedged, he can be removed through the nearest hoistway door.

Situation II: Victim Pinned Between the Car Floor and Hoistway Door Opening

This situation can occur when a car suddenly ascends while a person is exiting through the car doorway. (See Figure 3.14.) The following procedure is recommended for this type of situation:

I. Shut off all power to the elevator.

2. Stabilize the car by one or both of the following methods:

Method A - Use lash ropes or attach a lead line to the car top and/or bottom to limit movement.

Method B -Block or confine the car to stop movement using jacks, high-pressure spreaders, and/or wood props and wedges.

3. If possible, move the car to safely release the victim, under the direct supervision of an elevator mechanic.

Fig, 3.14. Mannequin's leg is pinned between car floor and hoistway door opening. A mattress has been used to block hoistway to prevent anyone from falling into it.

4. When there is any doubt about moving the car safely, cut the building material from around the victim. Block the

93

hoistway opening to prevent accidental falls before cutting. It is probably easiest to cut the hoistway enclosure wall near the elevator door. Then cut and pry the metal door frame to release the victim.

5. If neither of the methods outlined in the preceding Steps 3 and 4 is possible under the circumstances, cut away the car floor. This is usually more difficult and time consuming, and is best used only as a last resort.

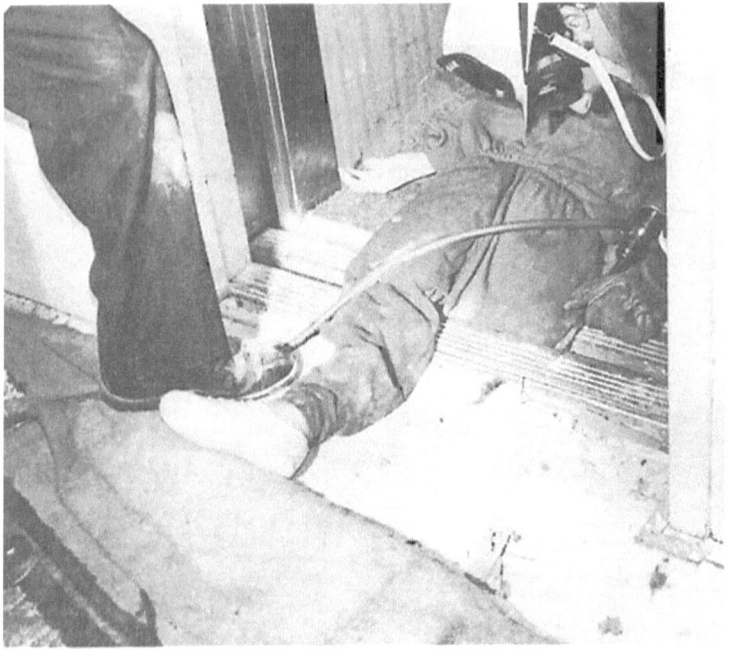

Fig. 3.15. *Mannequin is pinned between* car *floor and floor landing. Firefighters prepare to cut either floor landing or car floor away from* victim.

Situation III: Victim Pinned Between the Top of the Car Doorway and the Hoistway Floor Landing

This situation can occur when a car suddenly descends while a person is entering the car's doorway. The following procedure is recommended for this type of situation:

1. Shut off all power to the elevator.

94

2. Stabilize the car using one of the methods outlined in Situation II, Step 2.

3. Block the hoistway opening to prevent anyone from falling into the shaft.

4. Under the direct supervision of an elevator mechanic, move the car away from the victim if this can be done safely.

5. If this cannot be done safely, cut the top of the car and door frame to release the victim. If the floor of the building is made of wood, it may be easier to cut the floor near the victim.

6. If there is enough clearance, you may be able to remove the upper guide shoes on the car, as outlined earlier in this chapter under "Emergency Rescue of Pinned Victims." Using jacks and pry bars, force the car toward the back of the shaft to free the victim.

Situation IV; Victim Pinned Under the Car Floor Against the Hoistway Floor Landing

When faced with this situation, rescue personnel should use the following recommended procedure:

1. Shut off all power to the elevator.

2. Stabilize the car using one of the methods outlined in Situation II, Step 2.

3. Block the hoistway opening to prevent accidental falls.

4. Under the direct supervision of an elevator mechanic, move the car away from the victim if this can be done safely.

5. Cut either the hoistway or the car floor away from the victim, whichever is easier and safer. (See Figure 3.15.)

6. If both the car and hoistway floors are too difficult to cut, you may be able to cut the guide rails above and below the lower guide shoes and car safeties. (This procedure

was described earlier under "Emergency Rescue of Pinned Victims.")

AN ILLUSTRATIVE EMERGENCY

The following incident which occurred in suburban Chicago illustrates the unique nature of elevator emergencies; it also illustrates that the fire service needs to be prepared for almost any type of emergency rescue situation.

One spring morning, a call came into a fire department that a man was trapped in a hotel elevator on the fifth floor. His right leg was pinned at the thigh between the elevator car floor and the fifth floor landing. An elevator malfunction had caused him to fall against the doors between the second and third floors, and his right leg had been forced through the doors into a three-inch opening between the car and hoistway wall as the car traveled from the second to the fifth floor.

Responding emergency rescue personnel made several unsuccessful attempts to widen the gap between the elevator car floor and the fifth floor landing with gas powered tools. An elevator service mechanic at the scene unsuccessfully tried to release the car, whose doors were in a semi closed position. When a second engine company arrived on the scene, they unsuccessfully tried to release the doors with one more power spreader. The three-inch opening had been widened to about ten inches, but the victim's leg was still pinned. Two firefighters were holding the victim so he would not slip down into the opening, when suddenly he went limp. One of the rescuers shouted "He's free, he's free." The victim was quickly placed on a stretcher and rushed to the hospital.

The rescue, which took over an hour, involved the teamwork of nearly 25 people from an elevator company, fire and police departments, as well as several bystanders and hospital personnel. In summing up the experience, an inspector from the Fire Prevention Bureau observed; "Once again, I am reminded that what seems impossible is possible, will happen, and we had better be ready when it does."

FINAL PROCEDURES

When rescue procedures have been safely completed, close all hoistway doors and remove any tools used in the rescue. Place the elevator out of service by throwing the main line disconnect switch to the OFF position (if the removal was made with the power on). The appropriate building personnel should be delegated to take charge of the elevator to guard it against further use. Barricades should be put up where walls have been breached or hoistway doors cannot be locked (i.e., interlock secured) in the closed position.

SUMMARY

In any rescue situation, regardless of the degree of seriousness, passenger safety is the prime responsibility. Passengers need to be protected and reassured of rescue. Let them know what is being done for their removal. Don't take unnecessary chances remember, passengers are safe in the car. *Unless it is a dire emergency, wail for a qualified elevator mechanic to arrive on the scene.*

Following is a summary of the steps for *routine* elevator rescue. In most situations, steps five, six, seven, or eight will free passengers.

1. Contact a qualified elevator mechanic or contact the elevator company.

2. Contact the passengers to reassure them and learn their condition.

3. Locate the car.

4. Check the power supply system(s) and turn it off and on twice.

5. Have a passenger engage and disengage the Emergency Stop Switch.

6. Push the landing button while a passenger pushes the Door Open or Floor Select button.

7. Shake the hoistway and elevator doors.

8. Try to break the light beam.

9. Turn off all power supply systems.

10. Have a passenger open inner door.

11. Have a passenger open outer (hoistway) door by tripping the interlock, if possible.

12. Release interlock yourself.

13. Pry the door open.

It is advisable to determine what steps your department would take during an elevator emergency by studying the elevators in your response area. Without prior knowledge and familiarization with your community's elevators, there is no telling whether the rescue procedures discussed in this chapter will work for you. It is ~1:andard good practice to maintain contacts with the elevator mechanics and manufacturers in your community. Their knowledge of elevator systems will be a great help to your department in formulating procedures for handling emergencies.

CHAPTER 4

Emergency Use of Elevators

INTRODUCTION

Elevators enable high-rise buildings to exist: without them it would be impossible to move so many people such great vertical distances. The equipment that runs these machines is of necessity sophisticated and complex; under normal conditions, it provides service unparalleled for safety and efficiency.

Under abnormal conditions, however, the situation may be quite different. Because elevators are electrically powered, exposure to heat and water can make them unreliable and even dangerous. Thus a fire in a building with elevators presents a triple threat. First, the normal use of elevators (the usual means of transporting occupants) is hazardous both for occupants and firefighters. Second, there are likely to be many people in the building, and their presence will complicate fire fighting. Finally, because of the building's height, the fire itself may be hard to reach. The sum of this is that fire operations may be very complicated, and it is essential that thorough preplans be drawn up for each large building. This chapter will explain how elevators may be used to advantage in fire fighting. Chapters Five and Six will discuss machine room fires and problems of evacuation.

PREPLANNING

Elevators vary from building to building, manufacturer to manufacturer, and year to year; they are also subject to the changing requirements of various local, state, and national codes. For this reason, it is impossible to overemphasize the importance of preplanning. Nearly every elevator installation is different in *some* way from others, and ignorance of the particulars can have serious or even fatal consequences.

While this chapter will explain *in general* how to use elevators in emergencies, it is impossible to make all-inclusive statements: almost no single piece of information applies in every case. It is suggested, therefore, that this information be used to form general guidelines and principles that may be modified to fit particular circumstances. The pre-planner should know all of the codes -local, state, and national - that apply to elevators in his or her area, and be aware of changes in both codes and installations. It is also important *to* establish good working relationships with building owners and elevator mechanics. Knowing elevator mechanics is particularly important, for there is no one who knows more about the idiosyncrasies of individual systems. It should be determined which buildings in an area have elevator mechanics on duty at all times and, in those that do not, how a mechanic can be reached in an emergency. The mechanic should be called not only during an elevator rescue but also during a general emergency such as a fire. The presence of a trained mechanic in such situations might avert disaster.

Because codes, buildings, and installations differ so widely, no book can cover every variation. Rather than try to, it is better simply to repeat: *there is no substitute for preplanning.*

THE FIREFIGHTERS SERVICE

Many firefighters have found that their greatest ally when fighting high-rise fires is the Firefighters Service. Also called the Emergency Service, Fireman's Switch, or some similar term, it is an emergency feature designed by elevator companies to reduce the risks of using elevators during fires. (It may also be useful in a few other situations: during peak use periods for example, firefighters might use it to reach a medical emergency on an upper floor.)

The Firefighters Service can be of immeasurable help in fire operations, but it must be used properly. Because details of use vary from building to building, and understanding the basic purpose of the system is important, it seems better first to discuss the Firefighters Service *out* of the context of fire fighting. In this way a more complete general idea can be formed before variations are discussed. The source of the current definition of the Firefighters Service is ANSI/ASME A17.1, *Safety Code for Elevators, Dumbwaiters, Escalators, and Moving Walks,* and all information in this section derives from code requirements.

The Firefighters Service is basically a two-phase system: Phase One captures all elevators automatically or manually and brings them to the ground floor of a building during a fire. Phase Two allows firefighters to run elevator cars independently of automatic service in order to use them to reach upper floors. ANSI/ASME A17.1 requires any elevator constructed after 1973 to have Phase One if it serves at least three landings or travels 25 feet; it requires both Phase One and Phase Two in elevators built after 1973 which travel over 70 feet or have a terminal landing 70 feet above the ground. In other words, only Phase One (car capture) is required in low-rise buildings, since a fire on an upper floor could be reached easily. In taller buildings, however, firefighters must be able both to capture cars and use them to fight fires. One can readily see that the logistics of fighting a fire on the thirtieth floor could become extremely complicated if each firefighter had to climb all thirty stories to reach the fire. In addition to describing *where* the Firefighters Service should be installed, ANSI/ASME A17.1 also specifies *how* it must operate on elevators constructed after 1973. In practice, of course, this varies, but the following summarizes the ANSI/ASME Code's requirements for Phase One:

- At the main floor, a three-position (ON, OFF, BYPASS) switch must be provided for each elevator or bank of elevators. (See Figure 4.1.) The keys that operate the system must be identical, must not be part of any master key system the building may have, and must not be removable when in the BYPASS position. The key must be readily available to emergency personnel, and there must be a key for each three-position switch.
- When the switch is in the ON position, all elevators that are controlled by this switch *and* that are on automatic service will return without stopping to the main floor. Upon reaching the main floor, the doors will open and remain open. Any elevator

101

traveling away from the ground floor will reverse at the next available floor without opening its doors and proceed at once to the main floor. An elevator whose doors are open on a floor other than the main floor will close its doors immediately and return at once to the main floor.

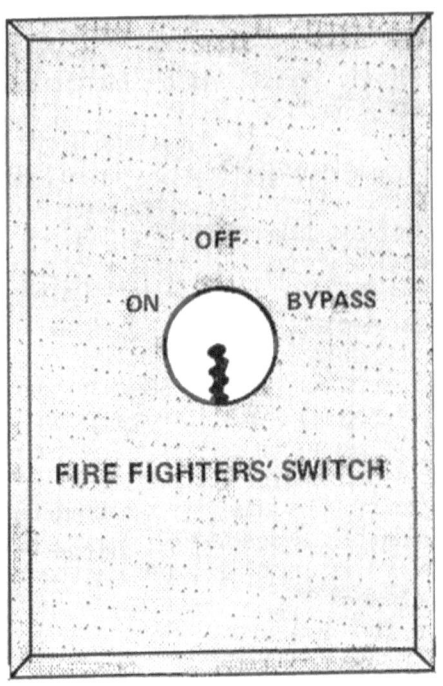

*Fig. 4_1. By operating the Lobby Key Switch, firefighters can capture elevator cars on automatic service and bring them nonstop to the ground floor during a fire_*

- The Emergency Stop Switch of a car with open doors will cease functioning as soon as the car doors are closed and it begins its descent. The Emergency Stop Switch of all moving cars will cease to function immediately and will remain that way until the cars reach the main lobby.

- Heat and smoke detectors are required in the elevator lobbies of each floor, with these exceptions:

    (1) the main floor,

102

(2) buildings completely protected by automatic sprinklers,

(3) freight elevators in manufacturing areas, or opening onto manufacturing areas,

(4) elevator lobbies with unenclosed landings. If these detectors sense a fire, they will automatically activate Phase One of the Firefighters Service and bring all elevators to the main floor in the previously described manner.

- Turning the key to the BYPASS position cancels any smoke detectors or sensing devices and restores elevators to normal service.

- Operating instructions for the system should be posted adjacent to the switch at the main floor, and should explain Phase One and Phase Two (if provided). Letters should be *Y4 U* high and signs should be permanently installed and protected against removal and defacement.

Phase One of the Firefighters Service is required on nearly all elevators now built, and its purpose is to remove elevators from normal service should there be a fire. Note that it may be either activated *automatically* (by smoke or heat detectors) or *manually* (by someone with a key).

The purpose of Phase Two is to enable firefighters to use elevators in fire fighting. All of the automatic features canceled in Phase One remain inoperative; i.e., cars will not respond to hall calls, automatic door close features will cease functioning, etc. However, the car can now be put into use. Following is a summary of the ANSI/ASME Code's provisions for Phase Two:

- The Lobby Key Switch must remain in the ON position during all Phase Two operations.

- Once the Lobby Key Switch has been activated and all cars returned to the ground floor, a single car or cars may be operated. Any car capable of Phase Two of the Firefighters Service will have, plainly visible, a key switch *inside* the car. This switch must be turned to ON.

• Automatic features of the elevator will not operate. To operate the elevator, someone must register a floor call and, in some elevators, push the Door Close button in order to close the doors. The car will then travel to the chosen floor and stop. To open the doors, someone must press the Door Open button. This button requires continuous pressure until the doors are *fully* open; if it is prematurely released, the elevator doors will close.

VARIATIONS IN THE FIREFIGHTERS SERVICE

As noted before, a multitude of variations in the Firefighters Service occur in the field. These variations are mostly in buildings built prior to 1973 that have since installed the system, and in places where local codes specify something other than what is in ANSI/ASME A17.1. Since this system is extremely expensive to install in existing buildings, local boards have sometimes granted variances. In some buildings, for example, the Emergency Stop Switch may not be cancelled by the Firefighters Service. In other buildings, particularly older ones, the three-position switch for a bank of elevators may be only a two-position switch. (See Figure 4.2.) This would be the case if the building has no smoke-detecting system tied into the elevator installation; there is then no need for a BYPASS position, because the Firefighters Service can only be activated manually, not automatically. Hence, if the Lobby Key Switch is at OFF, elevators run normally. In systems tied into smoke detectors, however, the Firefighters Service may be activated without a key; in order to run elevators normally a BYPASS position is necessary.

The possibilities for variation seem to be endless. In many areas, Phase One is activated not only by smoke and heat detectors but also by water flow in the sprinkler system or someone pulling a fire alarm. Elevator officials object to this, claiming that it takes elevators out of service unnecessarily. They point out that a small fire in a back utility closet might be easily extinguished by a sprinkler and pose no threat to elevator users. If Phase One is tied into sprinklers, then elevators will be captured unnecessarily when the sprinkler activates. But fire officials have pushed to have all alarms tied into Phase One; they say that no matter what the size or location of the fire, elevators should be ready and waiting for them when they arrive.

Fig. 4.2. The Lobby Key Switch is often a two-position switch. This is the case if a building does not have an automatic capture system and there is no need for a "Bypass" position.

Other variations may be found in the appearance of the Firefighters Service and the number of cars capable of Phase Two. In one building examined during research for this book, the Firefighters Key Switch for elevator banks was not on the banks themselves (as specified) but across the lobby on the main control panel. (See Figure 4.3.) In some installations, there may be an audible signal or other indication that the Firefighters Service has been activated and the cars captured (Phase One). Some buildings have red signals that light up over any cars that may be used for Phase Two. Even in buildings where Phase Two is required, it may not be required on all cars. In Massachusetts, for instance, if an installation has more than two cars, only 50 percent of all cars need to be capable of Phase Two. In California, all cars in an installation must be capable of Phase Two; in addition, one car *out* of normal traffic patterns in a separate hoistway must also be provided.

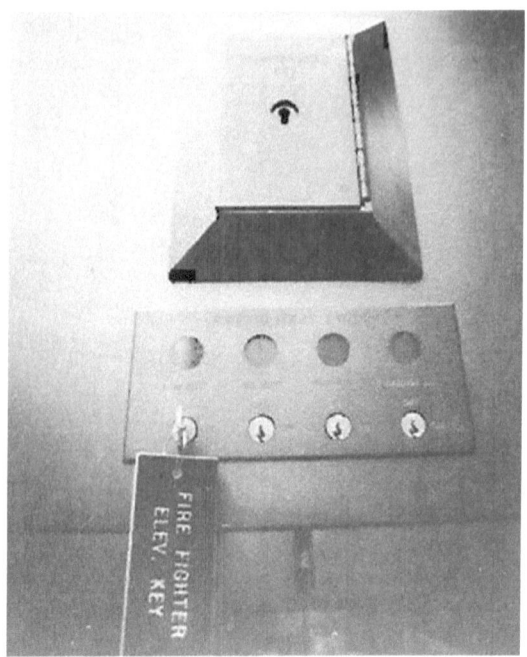

Fig. 4.3. *Occasionally the Lobby Key Switches are located not on the banks of elevators, but all together at the Lobby Control Panel.*

There are also many variations with regard to actual car use, such as how to hold doors open, etc.; some of them will be discussed in the section called PROCEDURES. Even if the fire service could somehow enforce ANSI/ASME A17.1 consistently, there would still be variations, for federal government buildings are not subject to local codes and are seldom built according to them. Once again, the statement bears repeating: *there is no substitute for preplanning.*

PROCEDURES: CAPTURING CARS

The remainder of this chapter will deal, in rough chronological order, with steps firefighters should take at a fire in a building with elevators. Immediately upon arrival, firefighters should capture all elevator cars and take them out of service. This must be done whether or not they will be used later in fire fighting, because automatic elevators are particularly vulnerable to heat, water, and smoke. Intense heat may, for example, short-circuit a hall call button and register a false call. If a car answers this call, any passengers within will be exposed to the fire. Moreover, if the car has a photoelectric door closing device (as many do), smoke from the fire may break the light beam and keep the doors from closing once they have opened. Water running down the hoistway is often a cause short-circuits in electrical switches and wiring.

Therefore, elevators must be taken out of service *immediately* if there is a fire. Car capture consists of bringing all cars to one floor and grounding them. Generally there is only one floor from which cars may be captured; it is usually the first floor, though for some freight elevators it may be the basement. Car capture is most easily accomplished if the building has Firefighters Service.

Capturing Cars With Firefighters Service: All firefighters and officers should be familiar with the Firefighters Service and, if possible, the variations that occur in it locally. Each building with the Firefighters Service should have a locked repository box containing the keys necessary in an emergency. These include a key for each bank of elevators and a key for each elevator able to be run independently under Phase Two. They might also include hoistway interlock release keys, if such exist. The repository box should be prominently located and marked; often it is red, with white letters printed on the outside that say "Firefighters Key." (See Figure 4.4.) The repository box key should be the same everywhere in a jurisdiction, and every company officer should carry one. Because repository boxes are sometimes vandalized, many cities either have the box protected by the building's security system, or provide company officers with the key that actually operates the Firefighters Service. As with the key to the repository box, the keys that activate the Firefighters Key Switch should be the same throughout a jurisdiction. Some fire departments also ask that the name and number of the elevator company be posted on the repository box so that a mechanic can be contacted quickly.

Fig. 4.4. The repository box should be prominently placed and should contain all keys necessary to firefighters in an emergency.

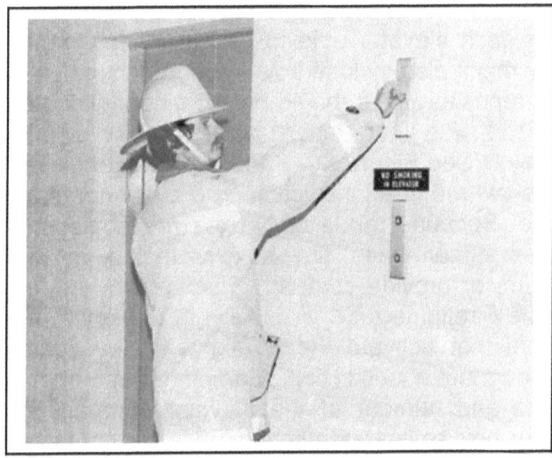

Fig. 4.5. If a building has the Firefighters Service but does not have an automatic capture system, firefighters must capture cars manually by turning the Lobby Key Switch to ON.

Upon arrival at the scene of a reported fire, officers should immediately capture all elevator cars. If there is an automatic capture system tied into the Firefighters Service, the elevators will have already returned to the main floor. (N.B.: Many buildings are designed so that when a fire occurs on the main floor, the elevators will proceed to a designated alternate floor. This is true in almost all federal buildings.) If there is not an automatic smoke detection system, an officer must obtain a key for each bank of elevators and manually activate the system. (See Figure 4.5.) As this is being done, a firefighter or prerecorded message should explain to passengers via the intercom what is happening. (Usually there is an "all call" button which allows the speaker to reach all cars at once.) Otherwise they may panic and set the Emergency Stop Switch. In some installations, Phase One of the Firefighters Service does not void the Emergency Stop Switch and the car will stop at once. (Where this is the case, it is recommended that the building owner correct it.) In a burning building, an elevator suspended in midshaft with panicky passengers could create an extra emergency. At any rate, passengers should be informed, if at all possible, when cars are being brought nonstop to the main floor.

Capturing Cars Without Firefighters Service: Because of Code requirements, nearly every new elevator installation will be capable of Phase One. But most older buildings do not have this feature; because it is so expensive, they are unlikely to install it unless required to by law. With these buildings preplanning is particularly important, for wide variations are likely to exist. In some cases, there will be central controls on the main floor located behind a locked door or panel. This door or panel can be easily pried open if necessary. The controls inside are normally used by elevator mechanics during maintenance and repair work, and one of them will usually be a switch capable of calling elevator cars to the main floor, (In some cases, cars may first have to complete their cycle of calls.) This switch should be activated. If at all possible, inform passengers of what is taking place because in all elevators without Firefighters Service, the Emergency Stop Switch will continue to function. If a passenger sets this switch, the car will stop in midshaft and create a major complication,

Fig. 4.6. The Firefighters Service will not capture passenger service cars operating out of automatic service. An audible signal and a sign identifying it should be arranged so that the car will be brought to the ground floor if it is needed.

In a building where there is no Lobby Control Panel, use the intercom (if one is available) to communicate with passengers. Tell them to enter a lobby call on the car panel. Send a firefighter to the machine room with a two-way radio and then push the Up button at the Lobby Elevator Call Station. As soon as an elevator comes to the lobby, set its Emergency Stop Switch at OFF

109

and have the firefighter in the machine room throw the main line disconnect for that elevator to OFF. Repeat this procedure until all elevators have been removed from service. It is important that the main line disconnect be thrown for each elevator after it is returned to the lobby; otherwise, it may not be possible to call the other elevators to the lobby.

Capturing Service Cars: Elevator manufacturers distinguish between true freight elevators and passenger service cars. Many buildings have both types of elevators in addition to passenger cars. Freight elevators are generally used *only* for carrying the operator and freight handlers, whereas passenger service cars are used by building maintenance personnel. In buildings with both kinds of cars, the passenger service cars are almost always tied into the Firefighters Service, Phase Two as well as Phase One. (Sometimes the passenger service cars have a separate Lobby Key Switch, but more often they are tied into the High-Rise Group for Phase One.)

As the passenger service car is very often taken out of automatic service by building personnel, it will not always be captured by the Lobby Key Switch of the Firefighters Service. Therefore, a means of signaling the operator must be devised. Pushing a button adjacent to the Lobby Key Switch or (in some cases) activation of the Lobby Key Switch alone will usually sound a bell on the roof of the elevator; this signals the operator to return the car immediately to the lobby for fire department use. Signs in the cars themselves should explain what the sound signals. (See Figure 4.6.) It has been learned from experience that a very loud bell or klaxon horn is necessary for those periods when the elevator operator has left the service car. Whatever steps are necessary should be taken to ensure that the passenger service car will be returned to the lobby during a fire. It is especially valuable in buildings with more than one bank of elevators, since it: (a) is larger, (b) is capable of stopping at *all* floors, and (c) eliminates the possibility of becoming stranded in a blind hoistway.

PROEDURES: USING ELEVATORS TO FIGHT FIRES

Using elevators to fight fires requires extreme caution. If a fire is on a lower floor, it is best to avoid their use altogether. But there are times when using elevators is helpful and even necessary. For example, if there is a fire on the thirtieth floor of a building, relief crews will be needed earlier and more often if every firefighter has to climb all thirty stories to reach the fire. Keeping track of this increased number of personnel will make operations more complicated and difficult. In addition, firefighters climbing stairs will be traveling in the opposite direction from occupants who are being evacuated, creating even more confusion. In cases like this the careful use of elevators may greatly expedite fire operations.

If possible, firefighters should only use elevators with Phase Two of the Firefighters Service in fire operations. If it is absolutely necessary, elevators without Phase Two can be put on independent service and used, though this is not recommended. At any rate, automatic elevators should *never* be used in fire fighting; as noted before, they are extremely vulnerable to the effects of smoke and heat. Even if all firefighters have to climb thirty stories to reach the fire, it is better to do that than risk lives by using elevators on automatic service.

Operating a Car on Firefighters Service: In a building which does not have automatic car capture, Phase One must be activated manually by turning the Lobby Key Switch to ON. If the cars have been captured automatically, the Lobby Key Switch should be turned to ON to give visual proof that the system is in operation. Once this is accomplished, it can be determined which cars are capable of Phase Two. Any cars capable of Phase Two will have a Car Key Switch on the control panel inside the car; this switch will be labeled Firefighters Switch or something similar, and will use the same key as the Lobby Key Switch. Once the proper key has been inserted in this switch and turned to ON, the car should be able to be run on Firefighters Service.

There are several common types of car switches. Most common is the two-position switch shown in Figure 4.7. In this model, the key is simply inserted and turned to ON. Once the switch is turned to

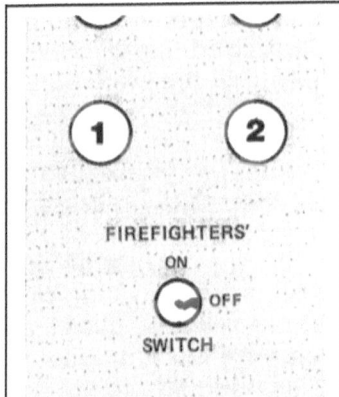

Fig. 4.7. The Car *Key Switch enables firefighters to operate an elevator car independently and use it to reach upper floors during* a *fire.*

Fig. 4.8. Some cars *have* a *separate Door Switch which must be set in order for the* car's *doors to stay open.*

ON, the key mayor may not be able to be removed, depending on the installation. A variation of this is a center-removal switch that fulfills the same basic function, but is operated differently. In this type, the key is inserted and turned to ON (roughly equivalent to 2 0'clock). However, the key cannot be removed unless the switch is turned first to the 12 o'clock position. To remove the car from Firefighters Service and return it to normal operation, the switch must be turned back to 10 o'clock and then returned to 12 o'clock for key removal. This second type of switch is not now installed, because it is difficult to know at a glance whether the Firefighters Service is in use or not.

A third type of car switch is the three-position switch specified by an older version of ANSI/ASME A17.1. This switch is present in installations with smoke detection systems, and has positions for ON, OFF, and BYPASS. In this type of car switch the Firefighters Service is on when the key is turned to ON, and off when the key is turned to OFF. When the key is turned to BYPASS, the elevators will bypass the door interlocks. This feature was found to be extremely dangerous; where it exists, it is recommended that the owner remove it.

Though the purpose of Phase Two of the Firefighters Service is the same everywhere -to enable cars to be run independently to fight a fire -the way that cars actually function may be very different. In all installations, activating the car switch should void all automatic functions, including response to hall calls, door-opening and closing devices, and photoelectric cell. The differences arise in how the car functions independently. In some installations, someone must press the Door Open button and keep it pressed until the doors are fully open and the Emergency Stop Switch is set in order for the doors to stay open. In other installations, it is only necessary to operate the Door Open button till the doors have fully retracted and remained in this position for a moment; after this they will stay open until Door Close is pushed. In still other cases, there is a toggle switch labeled Firefighters Door Switch next to the car switch (see Figure 4.8). If there is a toggle switch it must be moved to ON in order for the doors to remain open. (N.B.: If there is a toggle switch and a careless passenger or child pushes it to ON, automatic service will not be affected. But should it be necessary to activate the Firefighters Service, Phase Two *will not junction* in some installations unless the toggle switch is returned to the OFF position.)

Once again, there is no substitute for preplanning. In general however, the following explains how Phase Two of the Firefighters Service operates.

- Activate Lobby Key Switch. Enter car, insert key, and turn Car Key Switch to ON. This will void all automatic features.

- To close door, operate Door Close button and hold it until doors are completely closed, or enter a call.

- Push floor selector for desired floor, and Up directional indicator, if there is one. 'When elevator reaches this floor, it will stop. The doors will *not* open (as a protection from smoke, fire, etc.) .

113

- To open doors, push the Door Open button. Releasing this button before doors are *fully open* will allow the doors to reclose. In many cases the doors will be held open once they have retracted fully. In other installations, setting the Emergency Stop Switch or a special Firefighters Door Switch will hold the doors open. In *no case!>* will they remain open if the Door Open button is released prior to full retraction.

- Make sure that one firefighter remains with the elevator and maintains communication with the main lobby or Fire

Fig. 4.9. If cars must be used in buildings without Phase Two, firefighters should open the locked car panel and void automatic services. Here the button marked NS is the Independent Service button.

Command Center by means of the elevator's intercom or portable radio. If the elevator doors close without an operator present, reentry into the car will be impossible except by methods used to gain entry to a stalled elevator.

- To travel to another floor or return to the lobby, enter the appropriate call. On some elevators this is all that is necessary to resume travel, but on others the Door Close button must be pushed and held until the doors are fully closed before the elevator will start. A few installations have a Reverse Down or Reverse Travel button which it is also necessary to push. Otherwise the car will travel to the farthest point in the direction it has been traveling before reversing direction and stopping at the desired floor or lobby.

- If you realize that you have registered a call that will take you to or past the fire floor, turn the Firefighters Key Switch to OFF. Leave the switch in this position until the car returns automatically to the lobby and the doors fully open.

Operating a Car Without Firefighters Service Independently: If an installation does not have Phase Two of the Firefighters Service, it will be necessary to switch elevator cars from automatic to independent service. Almost all elevators have an "Independent Inspection" or "Hospital Service" that removes the car from normal group operation. Under normal conditions, this feature might be used by a mover, nurse, or elevator mechanic. In each car there will be a locked panel, which should be unlocked or, if necessary, pried. Behind the panel there will be a number of controls, as shown in Figure 4.9. The control labeled "Independent Service," "Ind. Ser.," or "I.S." should be operated, and any automatic door opening or closing device should be shut off. (Caution: Do not activate any switch marked M.G., as this will put the car out of service.) Once this is accomplished, the car will be able to be run much like a car on the Firefighters Service. Any variations in operation should be noted during preplanning.

PROCEDURES: GENERAL PRECAUTIONS

The preceding sections have explained the essentials of capturing elevator cars and operating them if this is necessary during a fire. This section will explain in larger context how elevators should be used. Much of this information is taken from *Fighting High-Rise*

Fig. 4.10. Here a firefighter is raising the top escape hatch to check the hoistway for smoke. On newer installations, this will not be possible because the top escape hatch will be bolted from the outside.

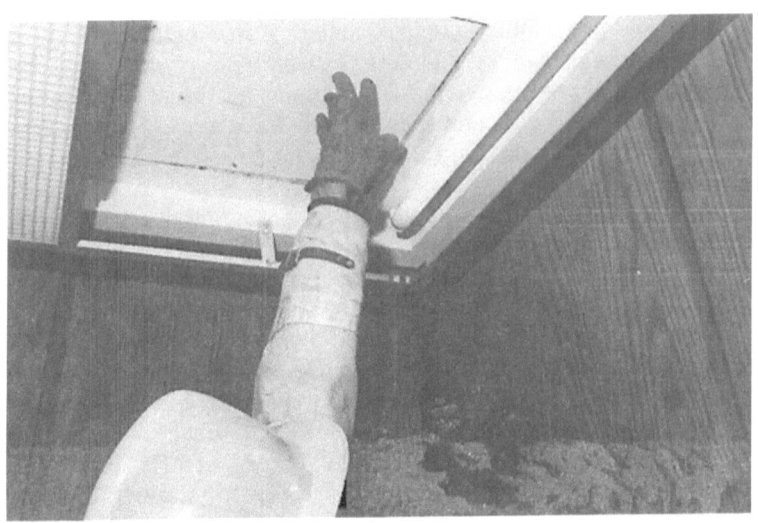

Building Fires -Tactics and Logistics, by Robert F. Mendes (NFPA, 1975). The following are general rules to remember when using elevators to fight fires:

- Most high-rise buildings have several banks of elevators. Elevators which serve the fire floor should be avoided if possible, as heat may render elevator behavior unpredictable.

- If the fire is below the 7th floor, avoid the use of elevators.

- If possible, use the passenger service car. This car serves all floors, SO caution is necessary. But it is larger, and holds more men and equipment. It is also usually operated independently. (Caution: Some freight elevators do not have Phase Two of the Firefighters Service. In this case, they should only be used when there is no car with Phase Two available.)

- Know the layout of the fire floor. At the very least, inspect wall diagrams that show the nearest exit. Learn in advance where this exit is and how to reach it.

- Do not overcrowd elevators. Do not load elevators with more than six firefighters, and make certain that they have at least one Claw, Kelley, or Halligan tool. In an emergency, this tool may be very useful. All firefighters should wear self-contained breathing apparatus.

- Before ascending, check for smoke in the hoistway by shining a light up the hoistway between the car and the hoistway wall. Do not remove the top escape hatch, as it will usually have an electrical contact which will render the car inoperative once the hatch is opened.

- Maintain constant communication between the elevator car and the Lobby Command Post. Intercoms or telephones are required on all new elevators, and they can be used when available. Portable radios should be carried as an alternate means of communication; they also have the advantage of being wireless.

- Do not take the elevator directly to the floor from which the fire will be fought. Instead, test the elevator by stopping at intervals to make sure that the system is working. If the elevator behaves at all erratically, stop it and exit.

- Never take an elevator closer than two floors below the fire floor, or past the fire floor. If it is absolutely necessary to take a car past the fire floor, use an elevator that runs express past the fire floor.

- If an elevator car does not stop as desired:

  a) Turn off the Car Key Switch, and Phase One will return the car to the lobby.
  b) Set the Emergency Stop Switch, open the car doors, or open the side escape panel: if the building does not have the Firefighters Service, any of these should stop the car.
  c) Tell the Lobby Command Post what is happening; they may be able to stop the car remotely.
  d) Put breathing masks in place, and get near the floor. If the car stops at a fire floor, try to close the doors and travel to a lower floor, or if this fails, an upper floor. If all else fails, try to get to the predetermined exits.

- Keep water off the elevator floor and away from the hoistway; it can create electrical hazards. Perhaps the greatest damage to elevators during a fire is caused by water; it damages electrical components on the car's roof, panel, and underneath portion.

- Never use elevators if the hoistway or machine is on fire.

117

- Whenever possible, call a skilled elevator mechanic to the scene. He or she will know the best methods to use in an emergency.

ADDITIONAL INFORMATION

The following important but unrelated information concerns the use of elevators during fires.

Deactivating the System: Once the fire is out, elevators can be returned to normal service. This is usually accomplished by undoing what was done in order to capture cars and use them in fire fighting. The Firefighters Service may be deactivated by first turning the Car Key Switch to OFF and removing it, and then turning the Lobby Key Switch to OFF and removing it. In buildings where Phase One is activated by smoke and heat detectors, the detectors will have to be reset before elevators will run normally with the Lobby Key Switch at OFF. (If detectors are not reset and the switch is turned to OFF, cars will automatically be recaptured.) To run elevators on normal service *temporarily,* turn the Lobby Key Switch to BYPASS. In installations without the Firefighters Service, the automatic group service controls should be reactivated in both the cars and the lobby.

After any fire in which elevators are captured, an elevator mechanic should inspect elevators to make certain that they are in good working order.

Testing the Firefighters Service: In buildings with the Firefighters Service, the system should be tested frequently and thoroughly. Elevator equipment is very sensitive, and it is not uncommon to find dust accumulated on little-used contacts which prevents them from working properly. The barrels of key switches may also be rusty and difficult or impossible to turn. Though it is not the responsibility of the fire service to keep this equipment in good working order, problems may go unnoticed unless a fire official points them out.

Fires in Buildings Under Construction: Fires are more common in buildings when they are under construction than at any other time; they are usually the result of poor housekeeping or arson. Because construction sites differ widely, little can be said about them except that preplanning is essential. Of course there will be no emergency elevator system; the only functioning elevator is likely to be a hand operated hoist that receives power from a generator somewhere on the site. Fire officials should learn how this hoist is operated and obtain from construction officials any keys that would be needed if a fire occurred at night. For more information on preplanning, see Chapter 2, "Buildings Under Construction."

Capacitance Tube Elevator Buttons: Capacitance tube elevator buttons are flat buttons that register calls when touched; they are made only by one elevator company but are quite common. Believing that they are activated by body heat, some firefighters have worried that heat from a fire would register false calls. Capacitance tube buttons are *not* activated by heat; they are activated by the static electricity from a person's fingertip. False calls are registered only after very intense heat has short-circuited the mechanism, a phenomenon that occurs with nearly all elevator buttons.

Delays in the Use of the Firefighters Service: Although the Firefighters Service is designed to be safe and efficient, there are cases in which a delay in its use may be dangerous. ANSI/ASME A17.1 now requires Phase One to be tied into smoke detectors, but there are many buildings without this feature. Instead they have a simple ON-OFF system which is activated by firefighters as soon as they arrive at the building. In these buildings) fire officials may want to consider training building personnel to activate Phase One, though elevator companies are generally opposed to allowing anyone besides firefighters to deal with the Firefighters Service. The reason is this: during the time between report of the fire and arrival of the fire department, elevator travel will still be possible, even though it might be quite dangerous. In a fire in a New York City office building, a terrorist's bomb exploded on the twentieth floor and caused a large fire immediately. In this fire the elevators were not damaged. But if a similar incident occurred in a building without Phase One tied into smoke detectors, elevator travel could become quite dangerous in a matter of seconds. If cars cannot be captured until firefighters arrive, the time lost could seriously complicate fire operations and endanger lives. Such a situation might be unusual, but then most firefighters have found that unusual situations occur far too frequently to ignore them.

REPORT Of A FIRE INVOLVING ELEVATORS

Beyond the general principles presented in this chapter, it is difficult to cover all situations. The following report of a fire in a large convention hotel offers an interesting illustration of this problem.

The fire occurred on the 29th floor, in the service elevator section at 1:08 a.m. This enclosed section has access to four elevator cars; they are referred to by number as shown in the sketch. (See Figure 4.11.) Lobby Call Buttons are the capacitance tube type: the call button is activated by an increase in the tube's capacitance when the static electricity in a person's body jumps from the fingertip to the metal in the button. (Sec Figure 4.12.) Car 9 is used by building personnel as a Fire Car. It responds to a call from an emergency switch in the fourth floor service lobby. This floor area is adjacent to the office of the chief engineer, the carpentry shop, etc.

The telephone operator reported a fire on the 29th floor to the two employees designated watch engineers. This was before there was any audible signal indicating a fire. The two employees left their office and made their way to the *service* elevator lobby nearby. One of the men summoned Car 9 (the fire car) by inserting a key in an emergency switch adjacent to the hall call button; the car came to the fourth floor. The men took the cart containing first aid fire fighting equipment into the car, entered the car themselves, and pushed the car's Floor Select button for the 29th floor.

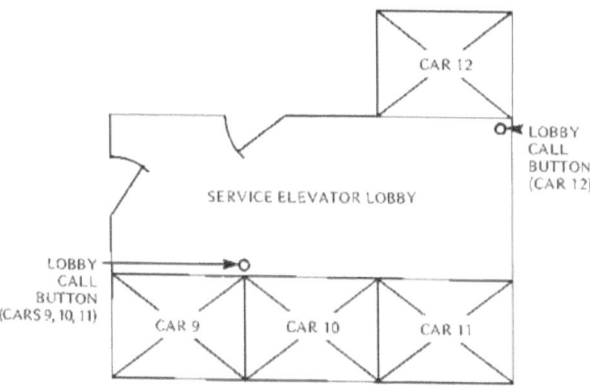

Fig. 4.11. The layout of the 28th floor service elevator section of a large hotel.

The occupants of the car said that when they arrived at the 29th floor, the elevator car doors tried to open but could not. They pushed a button for a lower floor, but the car failed to move. They then donned the filter-type masks that were part of the fire fighting equipment aboard the car. Later they were freed by firefighters who reached the area by using a different elevator bank in the guests' lobby.

The Chief Engineer, the two elevator repairmen, the officials of the maintaining elevator service center, several of the first arriving fire officers and firefighters, and one of the men trapped in Car 9 were all questioned during the investigation of the fire.

It appears what happened is that the hoistway doors of Car 9 became fused from the heat of the fire in the lobby. (The hoistway doors are coupled to the car doors, and each has an edging of a rubbery material. The edges come together when the landing doors close.) Although the doors could not open, they apparently parted sufficiently to break the electrical contact of the interlock, a mechanical latch which locks the two hoistway doors closed. Hence, Car 9 could not leave the floor because the interlock was unlocked. Firefighters using the Halligan Bar pried the doors apart with only a little effort, and found that the fire had done no damage to the hoistway doors.

121

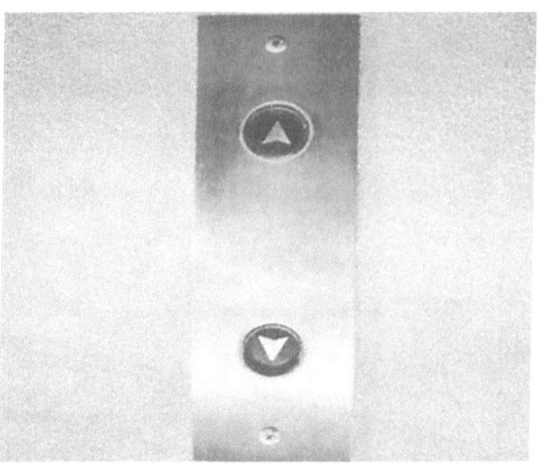

Fig. 4.12. Above are two capacitance tube buttons. The buttons affected by the fire were the same type, though not arranged in twos.

Car 12, the room service car, was found with the switch selector on the "attendant" position at the 28th floor. When a car is on "attendant'" service, it is removed from normal automatic operations: it is dependent upon the operator and will not respond to hall calls from the various floors. The use this switch position is common. It appears that the car was found at the 28th floor only because it was left there by the last person using it. However, the Lobby Call Button on Floor 29 which would be used to call this car when on automatic service was destroyed by the fire. It was distorted in such a way that it may have summoned the car to the fire floor without the "permission" of any person in the car. But since the car was not on automatic service, it did not respond. The distortion of the plate assembly and melting of the plastic call button were apparently caused by intense heal. (See Figure 4, 13.)

Cars 10 and 11 were on automatic service and were found at the fire floor. The Lobby Call Button on Floor 29 that controls these cars was damaged in almost exactly the same way as the other call button. It is concluded that intense heat short circuited the button and caused it to put in a continuous call from the 29th floor, which summoned Cars).0 and 11. Under normal conditions, the detent timer would automatically close the doors after a given interval and move the car to another floor. Because of the continuous firing of the defective button, however, every car on automatic service came to the floor . . . one at a time . . . in a perpetual sequence. The cars were then held at the fire floor, possibly because of warping of the doors from excessive heat, blowing of fuses, distortion from a fire stream, or even the melting of the same rubbery material which made the interlock inoperative in Car 9.

Firefighters report that they used a Halligan Bar to pry open the hoistway doors of both Cars 10 and 11, but that no heavy pressure was needed to accomplish this. Both cars were found empty by firefighters and, later in the same day, Car 11 was back in service. Firefighters investigating Car 10 found heavy burn damage to the car door's rubber safety edges when they opened the doors. This car was placed out of service for about ten days after the fire.

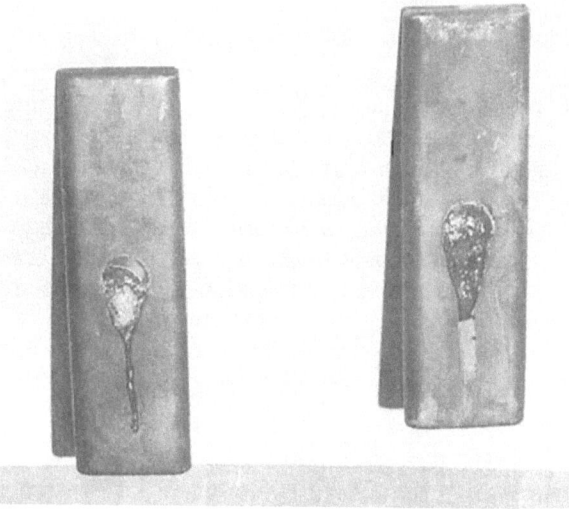

Fig. 4.13. Intense heat from the fire melted these two capacitance *tube call buttons and caused them* 10 *register false calls.*

123

CHAPTER 5

Elevator Installation Fires

INTRODUCTION

Most fires in elevator installations today are small and easy to extinguish. Formerly, serious fires usually occurred in hoistways and were the result of hazardous guide rails, traveling cables, and combustible construction; modern technology and codes requiring its application have all but eliminated these fires. Installation fires also develop in the machine room and the car itself-While fires in all of these locations are rare, ignorance of how to handle them could create a major emergency.

HOISTWAY FIRES

Most hoistway fires occur in the elevator pit; fires are more likely here because debris tends to accumulate and may begin to burn. Such fires are usually minor and in most cases may be put out with an extinguisher or short hose line. If the fire is in a hydraulic pit, however, it might present a Class B extinguishing problem, especially if the debris is floating in hydraulic fluid.[19] This fluid may burn intensely and create very thick smoke, which in turn may endanger passengers,

---

[19] Class B fires are fires in flammable liquids, gases, and greases; they must be extinguished with something other than water. Although tile oil used in hydraulic elevator systems has a high flash point, it may ignite if accumulated debris ignites first.

particularly if they are trapped. (See Figure 5.1.) Intense burning in the pit may short terminals or interlock circuits and thus stall elevators. If this is the case, any trapped passengers will have to be reassured and rescued as soon as possible.

Some elevator pits can be reached through the lowest hoistway door, which may be unlocked at its terminal landing with a key. Other pits may be reached through access doors. As these doors are usually locked, they will have to be either unlocked with a key or forced. Before entering the pit, engage the Stop Switch near the door (if there is one), and extinguish the fire through the open door or the door's vision panel (if there is one), (See Figures 5.2 and 5.3.) To do this, however, the elevator must be *above* the terminal landing.

When the fire is in the pit and the elevator door does not have a vision panel, the doors can be forced or opened with the interlock release key. Usually this must be accomplished by using a tool such

Fig. 5.1. The stains in this elevator pit are not the result of bad housekeeping but are normal occurrences. If debris and oil do accumulate, however, they pose a serious fire hazard.

Fig. 5.2. *The Stop Switch in* the *pit functions in the same way* as *the Emergency Stop Switch in the cars. This switch should be used as an extra precaution whenever possible.*

as a power jack. Apply force to open the door as described in Chapter 3. If the fire is extremely intense, it will be impossible for firefighters to enter the car and move it to an upper level, a procedure that is necessary when extinguishing a fire through a vision panel or open door. With hydraulic and electric traction installations, an elevator mechanic can run the car from the controller to raise the car. Otherwise, the hoistway wall below the car will have to be breached.

The fire hazard from lubricated guide rails has almost been eliminated with the introduction of nonflammable cleaning solvents and routine cleaning to eliminate grease and dirt buildup. At one time, traveling cables were a source of hoistway fires because they were likely to short out and burn. Today, codes require traveling cables to be flame retardant and moisture resistant. Should a traveling cable short, the problem may be corrected by opening the electric circuit. In older installations, the traveling cables may occasionally catch fire; these fires can be extinguished from the side exit door of an adjacent elevator or from a nearby hoistway door. Remove the electric power from the elevator before extinguishing the fire.

MACHINE ROOM FIRES

Although fires in elevator machine rooms are usually minor, they can present serious problems for trapped passengers under certain conditions. Since the machine room is usually built of noncombustible construction, it is unlikely to burn. However, maintenance personnel can and do ignore codes forbidding the storage of materials in spacious machine rooms, especially when they are located in the basement. Thus machine rooms may carry dangerous fire loads if they are used as storage areas for furniture, business records, supplies, etc.

When a fire starts in a basement machine room which is adjacent to the hoistway or elevator pit in installations having electric traction or drum elevators, the danger to passengers is especially great. Should this happen, the openings in the hoistway wall that separate the machine room from the hoistway will allow heat and smoke to move into the hoistway. The heat from such a fire could very possibly cause an electrical circuit to malfunction and thus endanger elevator passengers and expose them to heat and smoke. In such situations, firefighters must take quick action to rescue trapped passengers and at the same time fight the fire.

Electrical fires in the machine room could possibly endanger elevator passengers. These fires usually involve wiring, either for the building or the elevator's equipment. Electrical shorts usually trip the main circuit breaker and then burn out safely. However, the open circuit means that an elevator will stall and passengers will be trapped. In such cases, firefighters will also have to rescue trapped passengers and fight the fire simultaneously.

127

Fig. 5.3. If the door to the elevator pit has a vision panel-the panel can be broken and the fire extinguished through the hole.

Fighting an electrical fire involving elevator equipment requires the same methods and precautions used to fight any electrical fire. Before operations can begin, the power must be removed by throwing the main line disconnect switch. Because electrical fires usually generate dense smoke, firefighters must be very careful to avoid high-voltage electrical current, moving sheaves, and hoisting ropes while looking for the main line disconnect switch. (If smoke is very thick, it may be better to cut all power to the machine room at the building's main electrical panel.) In addition, firefighters should be careful not to fall through machine rooms which have secondary levels. If possible, building personnel should point out the disconnect switch or throw it themselves.

In general, fires in hydraulic elevator machine rooms are more intense than fires in electric traction installations. But since these machine rooms are blocked off from the hoistway, there is less danger to firefighters and trapped passengers. Moreover, the moving equipment of hydraulic elevators is usually enclosed in a single unit, eliminating the danger to firefighters from exposed elevator parts. However, combustible materials illegally stored in the machine room can catch fire and easily ignite leaking hydraulic fluid, especially if the fluid is not cleaned up regularly.

ELEVATOR CAR FIRES

Fires in elevator cars, though rare, are likely to be electrical and involve the lighting, fan, or car door opening motor. If this type of fire should occur, firefighters should locate the fused electrical disconnect switch or circuit breaker that controls the power for the lighting or fan motor and the main line disconnect switch for the elevator. Both switches should be turned to OFF. Power for the lights and fan motor normally comes from a source other than the elevator panel, and disconnecting the elevator's main line disconnect switch will not remove power for the light(s) or fan(s).

CHAPTER 6

Codes and Related Issues

INTRODUCTION

Thus far, this book has dealt with handling elevators during emergencies, both when the elevator itself is the problem, and when there is a more general emergency, such as a fire. This chapter will consider broader topics, including use of elevators as emergency exits and current provisions for elevator use by the handicapped.

USING ELEVATORS AS EMERGENCY EXITS

\Until recently, there was only one official attitude towards elevators during fires: "Stay out of them!" This applied to firefighters and building occupants alike; it was believed that any use of elevators should be avoided because the dangers associated with them were too great.

Lately, however, there has been evidence that this official view may be changing. The tactical problems of fighting fires (and the unpredictability of elevators) in high-rise buildings are such that the Firefighters Service has been developed and required on nearly all new elevators; in many areas, local ordinances now require that even older buildings install it. As a result, it is now possible to use elevators to fight fires, as explained in Chapter 4.

Today there are many people who would like elevators to be classified as supplemental emergency exits. This group is composed chiefly of building constructors, who seek a way to minimize rising construction costs, and advocates of handicapped rights, who claim that present codes discriminate against the handicapped. Together these groups are a formidable opponent of those who would keep building occupants out of elevators during fires. They maintain that the situation today is quite different from that of twenty years ago, and argue that codes should be revised to reflect increased knowledge and advanced technology.

Proponents of elevators as emergency exits emphasize that regardless of warnings, many occupants try to use elevators as exits during fires simply because people tend to leave a building the way they entered it. These proponents recall the Andraus and Joelma building fires in São Paulo; in the former, 200 people were trapped in the stairway, and in the latter, many of those who escaped did so via the elevators. These points are valid, but leave out some important information. Neither building in São Paulo had properly protected stairways, and in the Joelma building, one elevator operator died from the effects of the fire. Moreover, automatic capture is designed to keep people from using elevators during fires.

In most high-rise buildings, multiple protected stairways exist so that occupants can be evacuated to an area of refuge three or four floors away. In newer, larger buildings, elevators may playa part in evacuation. In an article published by the Society of Fire Protection Engineers, Stanley Kravontka notes the existence of megastructres like the Sears Tower in Chicago, the Empire State Building in New York, Bank of America Plaza in Atlanta, U.S. Bank Tower in Los Angeles, and Columbia Center in Seattle. When buildings reach heights like these, the use of stairways that run the height of the building is impractical during a fire. He quotes Asst. Chief Robert Burns:[20]

> .... The evacuation of occupant's more than six floors up or down probably is not feasible even for young healthy persons. Exit times for persons in high-rise buildings must be counted in hours, not minutes, if the building is large enough.

---

[20] Robert G. Burns, as quoted by Stanley Kravontka, "Elevator Use During Fires in Megastructures," (Boston: Society of Fire Protection Engineers, 1976) p. 4.

Fig. 6.1. The only emergency exit from most buildings ;s via the stairs, which many handicapped people are unable to use.

In buildings like these, protected stairways are still the primary means of egress. Once occupants reach an area of refuge, however, they may be evacuated by means of elevators under fire department supervision. Mr. Kravontka continues:[21]

> ......The future tower may not have stairs for evacuation from the tower as the primary function. The primary function of stairs would be as a means of egress from one hundred-foot cube to another, that is, from one cube with a fire condition to another cube for refuge.
>
> In the future tower, the elevators would operate within each hundred foot cube. Perhaps one freight elevator would operate over the entire height of the tower. This is substantially the operation in New York City's World Trade Center towers, each tower divided by so called "sky lobbies" into three separate buildings. The elevator travel is within these separate buildings, exclusive of a proposed

---

[21] 'Kravontka, p. 4.

freight elevator traveling the full height of a 1,200 foot rise and the feeder shuttles.

Elevators have not yet been approved as emergency exits, and there is still some question about whether they ever will be. For this to happen, the existing codes concerning exits and elevators would have to be changed, in particular the NFPA *Life Safety Code* . The *Life Safety Code* Committee has considered in the past allowing the use of elevators as emergency exits for the handicapped only, under the following conditions:

- Hoistways must be smoke-free and heat-free.

- The main lobby must be smoke-free and heat-free.

- Hoistways must be impervious to water.

- There must be available at all times both normal and standby power.

It is highly unlikely that these changes will be incorporated into the *Life Safety Code,* or indeed whether any changes will be made at all. If these changes were incorporated, it would be extremely costly to construct elevators that may be used *as* emergency exits. Although such changes would satisfy advocates of the handicapped, building constructors would most likely not be happy about the increased costs.

ELEVATORS AND THE HANDICAPPED

Elevators are usually considered liberators of the non-ambulatory since they free the handicapped from having to deal with stairs. But elevators *as* they now exist also pose a serious problem for the handicapped. Because people who cannot walk have to rely exclusively on elevators to travel vertically, the automatic grounding of elevators during a fire cuts off their only exit from a burning building. Until recently, this problem was not critical: the rarity of high-rise fires and the inaccessibility of buildings ensured that very few (if any) disabled persons were in danger. Now, however, federal law requires that buildings be made accessible. The non-ambulatory are living and working in these buildings with increasing frequency, and plans *must* be established for evacuating them if there is a fire.

Spokesmen for disabled rights groups point out that the definition of "non-ambulatory" is much broader in this case than some might think: it includes not only those in wheelchairs and on crutches but also, for example, the extremely fat and those with heart conditions. The latter might be able to walk for limited distances if the ground is level, but could never travel several flights of stairs to escape a fire.

What can be done to ensure the safety of the non-ambulatory during a fire? At present there is no one answer, nor does one seem likely. As explained in the previous section, many want to change the codes so that elevators can be used as emergency exits. But whether or not the codes are changed, most existing buildings are not likely to be modified because changes are too costly; evacuation plans for these buildings are needed. This issue is so new that many fire departments have not yet incorporated evacuation of the handicapped into preplanning, nor have elevator companies confronted it. In short, most handicapped persons now living and working in high-rise buildings depend upon the good will of their neighbors or coworkers to help them if there is a fire. The danger is probably not as great in an office building where people work close together, as in an apartment building, where doors are always locked and neighbors may hardly know one another.

Fig. 6.2. Even buildings with elevators may have stairs which render them inaccessible to the non-ambulatory.

Even halfway measures would be better than continued neglect of the problem. In particular, there are two courses of action that might improve the safety of the handicapped during a fire. Both of them, however, depend on building management knowing the location of the non-ambulatory, which may not always be possible. It is one thing to ask disabled persons living or working in a building to inform management of their presence and approximate location; but it is quite another to ask all non-ambulatory persons to register when they enter a building and sign out when they leave it, especially for short visits (as in the case of a salesperson calling on a customer). Neither the handicapped nor building personnel are likely to observe such a request; the handicapped will complain that it is restrictive, while building personnel will say that it is too complicated and time-consuming to administer.

As yet there is no sure way for building management to know exactly where disabled persons are in a building, but even partial knowledge is better than none. Should a fire occur, one of two courses of action can be taken. In the first, non-ambulatory persons (who have previously been told what to do) are moved to a designated place of refuge; they wait there for firefighters, who will rescue them either by using Phase Two of the Firefighters Service or by carrying them down the stairs. In the second plan, building personnel learn to use Phase Two of the Firefighters Service so that they themselves can evacuate the handicapped via the elevators.

Both plans present numerous problems. As for the first, what if the fire is in or near the area of refuge? Should the disabled merely move as far from the fire as possible and wait there for firefighters? How will firefighters know where to find them? And what if the building is too small to provide a suitable area of refuge? As for the second plan - teaching building personnel how to use the Firefighters Service -both fire departments and elevator companies strongly object to this. They know that the system is complicated and that using it requires many precautions; they are afraid that allowing building personnel to use it will create more problems than it will solve. Then too, what if the building is five stories high, and has only Phase One of the Firefighters Service? Should building personnel attempt to carry the disabled down the stairs, or should they wait for firefighters?

All of these objections are substantial, and emphasize how urgent it is for fire departments, elevator companies, building managements, and the disabled to discuss the problem of evacuation and work towards its solution. Probably there will not be one answer to the problem, but many: just as fire fighting strategy varies from building to building, so will evacuation plans. But whatever solutions are arrived at, the search for them should begin at once.

ELEVATOR CODES

There are codes which have sections dealing with elevators published by several organizations. Among them are the *Life Safety Code*, the *BOCA Basic Building Code*, the *National Building Code*, and the *International Building Code*. In addition, nearly all states and some cities have local codes which, within the boundaries of that city or state, carry the force of law. At some point almost all of these codes derive from or borrow from ANSI/ASME AI7.1, *The American National Standard Safety Code for Elevators, Dumbwaiters, Escalators, and Moving Stairs*. This code is published by the American Society of Mechanical Engineers, and was most recently revised in 2007. The ASME also publishes ANSI/ASME A17.2, an inspector's manual, to accompany the AI7.I Code.

If you are interested in elevator code sections as they relate to your jurisdiction it is recommended that you obtain a copy of the code book from one of the many suppliers or visit your local code enforcement office.

Glossary of Terms

ANSI/ASME *A17*.1: The official *American National Standard Safety Code for Elevators, Dumbwaiters, Escalators, and Moving Walks.* Covers design, construction, installation, operation, inspection, testing, maintenance, alteration, and repair of elevators, etc.

Buffer: A device which will stop a descending car or counterweight moving beyond its normal limit of travel by means of storing or absorbing and dissipating the kinetic energy of the car or counterweight. When a normal or final terminal stopping device fails, the car will stop on a buffer. Buffers are either oil or spring type.

Car Control Panel: Contains the signals used to select a desired floor, to show direction of car travel, and to close and open car doors.

Car Frame (Sling): The frame which supports the car platform. Guide shoes, car safeties, hoisting ropes or hoisting-rope sheaves are attached to the frame.

Controller: A device or set of devices programmed to control the apparatus to which it is connected by receiving all signals and dispatching elevator cars in answer to them.

Counterweight: On electric traction elevators and some drum-type elevators, a weight which travels up and down in the hoistway in its own guide rails, in the opposite direction of the car. Usually found at the back of hoistways.

Crosshead: The top beam of the car's frame.

Doors: The moveable portion of the car Or hoistway entrance which closes the opening from the car to the landing. Doors may be bi-parting, single slide, swing, center-opening, two-speed, or two-speed center opening.

Door Restrictor: A mechanical or electronic door restrictor that prevents elevator doors from being forced open outside a safe landing zone.

Drum Elevator: An elevator which operates by means of a spirally grooved drum onto which hoisting ropes wind as the drum turns.

Elevator: A mechanism for hoisting or lowering people and equipment in vertical structures. The mechanism consists of a car or platform which moves vertically in guides and serves two or more floors of a structure.

Electric -Traction Elevator: An elevator operated by means of power supplied by electric drive motors and consisting of an elevator car, counterweight, and traction sheave.

Hydraulic Elevator: An elevator operated by means of liquid under pressure contained in a cylinder equipped with a plunger or piston.

Elevator Car: The unit in an elevator installation used for carrying passengers or materials. Includes the platform, car frame, enclosure, and door.

Emergency Evacuation Bridge: A portable scaffold with guard rails which can be placed between the side exits of two elevator cars for safe rescue. The distance spanned should not exceed 2Y2 feet.

Emergency Stop Switch: A device located in the car which when manually operated removes electric power from the driving machine motor and brake of an electric elevator and from the electrically operated valves and/or pump motor of a hydraulic elevator.

Firefighters Service: (Also referred to as Emergency Service, Fireman's Switch.) The emergency feature designed by elevator companies to reduce the risk of using elevators during fires. (See Chapter 4, "Emergency Use of Elevators" for a detailed explanation of this feature.)

Firefighter's Service Control: (Also referred to as Firefighter's Key Switch.) Located on or beside the car's control panel, when activated by means of a key, this system places the car under the control of fire service personnel traveling in the car.

Governor Rope: A cable which passes over the sheave of the speed governor at the top of the hoistway and under a tension sheave at the bottom of the hoistway. It will trip safety switches at successively set speeds should an elevator begin to overspeed.

Guide Rails: Vertical steel tracks in the hoistway which keep the elevator car on a vertical path and minimize the car's side sway.

Guide Shoes (Roller Guides): Brake-type shoes used to maintain the elevator's vertical travel within the hoistway and connect the elevator to the guide rails.

Hall Call Button: The device, located at elevator landings, used by passengers to obtain elevator cars.

Hoisting Machine: For electric traction elevators, the traction hoisting machine which operates by the interaction of a traction sheave, driving motor, and motor brakes. For hydraulic elevators, the machine consists of a hydraulic pump, electric motor, and reservoir.

Hoistway: The vertical area of elevator travel. Also called hatch, hatchway, and elevator shaft. Can be used for one car (single hoistway) or more than one car (multiple hoistway). Includes all space from bottom of pit to underside of building roof.

Hoistway, Blind: The portion of a hoistway which passes floors or landings at which no normal landing entrances are provided.

Hoistway, Multiple: A hoistway providing for movement of more than one elevator car.

Hoistway, Single: A hoistway for one elevator car only.

Hoistway Enclosure: The vertical area set-off from adjacent areas of a building to enclose the elevator equipment within its own walls.

Interlock: An electro-mechanical device on the hoistway door which locks hoistway doors. Usually found on the header beam over the hoistway opening. Prevents elevator car from operating until it locks hoistway doors shut. Also prevents opening of the hoistway door from the landing side unless the car is within the landing zone and is either stopped or being stopped.

Interlock Release Key: The designated key which opens hoistway doors from the landing when inserted into the hoistway door unlocking keyway, if the door is equipped with the unlocking device.

Interlock Release Keyhole: A keyway found on the landing side of hoistway doors which are equipped with an interlock unlocking device.

Landing: The area of a floor used to receive and discharge passengers or freight.

Landing Zone: The area of the hoistway which extends from a point 18 inches below an elevator landing to a point 18 inches above the same landing.

Lobby Key Switch: A three-position (ON, OFF, BYPASS) switch located at the main floor and activated by firefighters using firefighter's service.

Lobby Supervisory Panel: A visual indicator of all elevator activity in a system, found in the lobby.

Machine Room: The room where the machinery used to operate any type of elevator system is contained. May be located in the basement, on the roof, or on the highest floor serviced by the elevator, depending on the installation.

Machine Room-Less: A relatively new elevator product termed Machine Room-Less Elevators (MRL's). All components are located in the hoistway. The Application is primarily in new construction.

Main Line Disconnect Switch: A fused-knife switch or large circuit breaker usually found inside the machine room near the entrance door. When thrown, it stops the car and removes all operating power from the elevator.

Motor Generator: Converts a building's alternating current to the direct current used by electric traction elevators.

Pit: The area of the hoistway which extends from the sill of the lowest landing to the hoistway floor.

Platform: Forms the floor of the car and directly supports the load.

Position Indicator: Shows the location of an elevator car in the hoistway. Called hall position indicator when located at a landing and car position indicator when located in the car.

Push-Button Station: The hallway or landing fixture used to call a car. May also indicate direction of movement.

Rated Load: The amount of weight that an elevator car is designed to lift at a rated speed.

Rated Speed: The speed, measured in feet per minute, at which a car is designed to operate with the rated load.

Safeties: The emergency brakes in an elevator installation which prevent the car from overspeeding. They are located at both ends of the safety plank and 9perate simultaneously on each guide rail.

Safety Plank: The bottom or floor joist area of an elevator car.

Selector: The device which starts, stops, opens, and closes elevator doors at designated floors.

Side Emergency Exit: An exit panel found on the side of the car's interior which, when unlocked, allows for escape from or entrance to another car in a multiple hoistway.

Speed Governor: A device located in either the machine room when it is above the hoistway or in the hoistway overhead which reads the speed of the car and activates elevator safeties during a malfunction.

Stop Switch: A passenger emergency signal located on the car panel which, when manually activated, will stop the car and signal an alarm.

Terminal Stopping Device (Normal): A device which slows down or stops an elevator near a terminal landing, independently of the usual stopping device.

Terminal Stopping Device (Final): A device which cuts off power to the driving motor and brake coil should the controller fail to stop a car normally.

Top Escape Hatch: An exit located on the top of an elevator car which opens outward and usually locks from the outside, or (on older installations) has a key or latch which can be opened from the inside.

Traction Machine; A direct-drive machine that operates an elevator car by the motion obtained through friction between suspension ropes and a traction sheave.

Traction Sheave: A grooved wheel that guides the wire ropes on an elevator.

Index

142

144

Instructional & Power Point Programs

The following Instructional and Power Point Programs (PP) are available for $49.95.

Terminology – Lesson Plan, Information Sheet, & PP (86 Slides)
Simple Rescue - Lesson Plan, Information Sheet, & PP (133 Slides)
Fire Operations - Lesson Plan, Information Sheet, & PP (67 Slides)
Complex Rescue - Lesson Plan, Information Sheet, & PP (41 Slides)

Send check or money order to the following address:

> STS
> Elevator Program
> PO Box 41
> Corvallis, MT 59828

Include your email address so the files may be sent to you electronically. Add $20.00 for shipping and handling if you would like a CD with the programs set to you via US Mail.

If you would like a to see a sample of the Power Point Programs email hmsk4mt@gmail.com under the subject "Elevator PP Sampler."

Web Sites

The following is a list of web sites that provide interesting information about elevators along with pictures and diagrams that may be of interest to emergency responders.

Elevator Bobs Web Site - http://www.elevatorbobs-elevator-pics.com/
Otis Elevator - http://www.otis.com/otis/1,1352,CLI1_RES1,00.html
ThyssenKrupp Elevator - http://www.thyssenkruppelevator.com/
Wikpedia - http://en.wikipedia.org/wiki/Elevator